PAINT CRAFT

by

TREVOR ROSSER

&

TOMMY SANDHAM

WILLOW PUBLISHING (Magor), Barecroft Common, Magor, Newport, Gwent, NP6 3EB, United Kingdom.

British Library Cataloguing-in-Publication Data.

Rosser, Trevor

Paint Craft

I. Title II. Sandham, Tommy

629. 26

ISBN 0 - 9512523 - 7 - 2

Production and Typesetting by Willow Publishing (Magor), Barecroft Common, Magor, Newport, Gwent, NP6 3EB, United Kingdom.

Cover design by A. R. Doe, The Studio, Llandevaud, Newport, Gwent, NP6 2AE, United Kingdom.

Printed in England by Morgan's Technical Books Ltd, P.O. Box 5, Wotton-Under-Edge, Glos, GL12 7BY, United Kingdom.

Contents Pages

Acknowledgements

This book could not have been completed without help from:

Bryan Moorcroft of the Lotus Cortina Register,

Trevor Rosser for his time and patience at the night classes, long before we ever decided to write a book,

Norman Paterson of Spray & Air Systems Ltd, Cwmbran, Newport, for advice when buying my spray gun,

Mike Pratt, Chairman of the MKI Cortina Owners Club, for giving me the honour of judging at their national rally in July 1993,

Barry and Mel Phillips of the Motor Discount Centre, Caldicot (where are you now?),

De Vilbiss company for supply of literature,

Hammerite Products Ltd for product literature,

Enthusiasm was also supplied by (and gratefully received from!):

Jo and Tony Doe for the front cover design, illustrations and general graphics help,

Garth and Jill Morgan for production and printing,

Bert Wiesfeld of The Netherlands, who came to the rescue for the second time with photographs taken during the paint spraying process,

Mike Phillips of MGP Restorations, for information about his Silurian,

Mrs. Sheila Taylor of The Magor Cellar for information about her MG,

Derek Thow for help propping up an ageing computer, (and an ageing operator),

My good friend ***Terry Burville*** for inspiration, criticism and a good all round sense of humour, plus his help in the two preparation chapters,

If I have missed anyone out I am sorry. The author records with gratitude the work put in by so many people to help this book towards publication.

Finally, to my wife ***Susan, and children, David and Rae*** -- thanks for putting up with all the problems, both real and imaginary, which occurred during the writing and production of ***Paint Craft.***

Warning

Paint Spraying can be dangerous.

Refer to the Health & Safety information on page 17, and the First Aid information below:-

First Aid

If Fumes are inhaled:

First, move the patient into fresh air. Keep the patient warm and still.

In very rare cases where breathing is irregular or has stopped, begin artificial respiration IMMEDIATELY. If possible, have someone call an ambulance. Continue artificial respiration until help arrives.

Do not give the patient anything to eat or drink.

When normal respiration starts, place the patient into the recovery position.

If paint material gets into eyes:

Wash eyes with lots of clean water for at least ten minutes. Keep the eyelids open during this irrigation.

Skin Contact:

Take off any contaminated clothing. Wash skin thoroughly with soap and water, or if available a recognised skin cleanser.

DO NOT use solvent or thinners to remove paint from skin.

Ingestion:

If paint materials are swallowed do NOT induce vomiting, but keep the patient still and seek medical attention.

Introduction

Up until 1992 I had never had much success with vehicle paint. Any attempts to paint anything always seemed to end in disaster -- like when I sprayed the front bonnet of a Ford Capri I was restoring. The Cellulose colour coat went on beautifully, and there were no runs. I stood back to admire my handiwork and immediately thought of putting on another coat, just to use up the paint remaining in the gun. Well, you can guess the next bit! The next coat went on top of the previous coat which had not yet dried, and the result was a mess of runs.

In 1992 I was restoring a MKI Cortina GT, and when the spraying estimates worked out at twice the cost of buying new equipment, I enrolled in Trevor Rosser's evening classes to learn to spray.

The complexity of the subject amazed me. Just as soon as you understood one bit of the subject, you peeled back one layer of the onion only to find lots more underneath.

I don't have a particularly good eye for colours or tones so this has been a bit of a handicap. However, I can now spray Cellulose and will get better (and more ambitious!) as time goes on. There is a tremendous satisfaction when you spray paint onto a panel and are happy with the result.

I still have to ask for advice from time to time but then, so does everyone else!

We have ensured that the information in ***"Paint Craft"*** is presented in a friendly, readable manner which will not put you off having a try at spray painting. We have not even mentioned paint chemistry anywhere!

Environmental agencies are focusing on the paint industry, particularly with regard to thinners and solvent emissions. This may change the approach to spray painting dramatically over the next few years. It is possible that the amateur paint sprayer, or more correctly the do-it-yourself paint sprayer, will become legislated against, and the freedom to do your own paint spraying may be taken away from you.

We believe that ***"Paint Craft"*** is the most comprehensive and up to date guide on the subject of paint and refinishing available in the U.K. today.

"Paint Craft" is a well-balanced work where the novice can be assured of finding a sympathetic approach, together with all the answers in one volume.

Good Luck, and I hope you enjoy your spray painting as much as I do.

Tommy Sandham,

Magor, Gwent,

September 1993.

Foreword

My first meeting with Tommy was when he attended one of my evening classes for basic spraying. The course consisted of the use of a spray-gun, spray-gun techniques, differences between paint materials, and fault finding. It covered a period of fifteen weeks and was aimed at students of any age.

Most of my students have never picked up a spray-gun in their lives, let alone used one!

After attending my class for twelve weeks, Tommy asked me if I had ever thought about writing a book on spraying. To say I was amazed would be an under-statement! To me, Tommy was just another student, but after a brief discussion I discovered he had his own publishing company and had already published several motoring and motor cycling books. He then told me that with my technical knowledge and his writing and publishing experience, we should be able to produce a book for the amateur sprayer -- which is what we set out to do.

Throughout *"Paint Craft"* Tommy has very cleverly taken my technical advice and used it by looking through his amateur sprayer's eyes and mind. He was very surprised by the amount of knowledge that is needed to become a successful sprayer. Just remember, that the more practice you get with a spray-gun, the better you become.

The main points to remember about spraying are:

1) Never rush a job,

2) Make sure your preparation is perfect,

3) Always ensure your spray-gun is spraying properly,

4) Always allow the paint to dry before carrying out the next stage,

5) If you do get a problem, always stop and work out what has caused it, otherwise it could cost a lot of time and money to correct.

Hopefully, we have achieved our aim and the contents of *"Paint Craft"* will help the amateur through the learning process.

Good luck with your spraying,

Trevor Rosser,

Newport, Gwent,

September 1993.

This Austin Healey 3000 is a work of art. Unfortunately, black and white photography does not do it justice. The upper half of the car is bright red, the lower colour being a rich creamy white. This paintwork -- even seen in black and white -- shows what can be achieved when you master the art of "PAINT CRAFT."

Glossary of Painting Terms

The following terms are frequently used when discussing paint spraying.

A working knowledge of these terms will aid your understanding of the paint spraying process.

A

Abrasive

A coarse material, (such as wet and dry production paper), used to remove paint or rust by rubbing it against the surface to be altered.

Adhesion

The degree of attachment between the paint coating and the surface or paint beneath it.

Air

The medium generally used to power a spray gun. Air is compressed in a compressor, may be stored in a receiver or holding tank, and is then forced along an air line to the spray gun.

Air Brush

A small air-fed spray tool used for detailed spraying or "customising" such as pin striping or lettering.

Air Hose

SEE Air Line, below.

Air Line

Also known as air hose. The rubber tube, often covered by a protective plastic coating, which is used to carry compressed air from the compressor to the spray gun.

Air Pressure

The amount by which air is compressed. It is measured in pounds per square inch (psi) or Kilonewtons per square centimetre (Kn/sq cm). In some circumstances pressure is measured in Bars, where one Bar (or Barometer) equals 14.2 pounds per square inch.

Air Pressure Drop

The amount by which the air pressure at the compressor is reduced when it reaches the spray gun. The pressure drop depends on three factors: the length of the air line, the internal bore or diameter of the air line, and the pressure at the compressor.

Air Receiver

Another word for an air storage tank or holding tank into which compressed air is pumped.

Atomisation

The process where the compressed air breaks up the paint and solvent into small droplets.

B

Binder

Binder is the agent used to keep the pigment in solution.

Bleed

The situation where one colour tends to show through another colour.

Blooming

A cloudy appearance in the finished paint surface caused by excessive moisture in the atmosphere when the spraying was carried out.

Glossary of Painting Terms

C

CFM

Cubic Feet Per Minute, the measure of the amount of compressed air passed through an air line.

Clear

A finishing coat, such as a coat of lacquer, without pigment (hence "clear") which protects or covers a colour coat.

Colour

The effect of light of varying wavelengths on the human eye. In paint colour is created by various tinted pigments.

Compressor

A mechanical device used to compress air. An electric motor (or a petrol or diesel engine) is used to drive a pump which usually consists of a cylinder and piston assembly. A Two Stage Compressor will have two such assemblies. Air is drawn into the cylinder and compressed by the piston. This compressed air is then either compressed again in the other cylinder or is passed to the storage tank or air receiver.

Coverage

The area which a given amount of paint will cover satisfactorily.

Crazing

Fine cracks on the surface of the paint. Can be caused by old age, or recoating a synthetic paint before the finish coat has dried properly. Also, an excessive delay in applying a second coat of synthetic where the first coat has started to cure.

Curing

The process of drying or hardening of a paint film.

Cutting

The use of T-Cut or compound to polish paint to a high gloss. See Polishing.

D

Double Header (Double Coat)

The process where two coats are sprayed without waiting for the first to "flash-off." Used to build up a thick layer of paint.

Driers

Substances which, when added in small proportions to oil-based paints results in appreciable reductions in their drying times at ordinary temperatures.

Dry Spray

Condition caused by holding the spray gun too far from the work. The compressed air tends to dry the paint too quickly giving rise to poor finish. Could also be caused by too high an air pressure.

Drying Time

The period between application of paint and a drying condition, determined by specified test conditions.

E

Elasticity

The flexibility of the paint film, much needed on plastics and fibre glass.

Enamel

A high gloss finish which dries slowly by evaporation of the solvent.

Etch primer

A primer which etches itself into a surface for good adhesion. Used on aluminium, fibre glass, metal etc.

Evaporation

The process where the solvent leaves the sprayed paint during the drying process.

Extenders

Extenders are an added property to allow the paint film to take longer to dry. Allowing the operator to cover larger areas without the fear of the paint film drying too quickly.

F

Feather-edging

The sanding process where a painted surface is worked until there is no step or lip where the paint meets the panel.

Filter

A filter is a device which removes contaminants from another material. For example lumps may be filtered out from paint, or water may be filtered out from the compressed air supply.

Fish-eyes

Small pits which will appear in the paint if silicone or wax has not been removed from the panel being sprayed.

Finish Coat

The top, gloss coat.

Flash-Off

The initial period when most of the solvent in the paint evaporates.

Flash Point

The lowest temperature at which a material, such as solvent, gives off sufficient vapour to ignite.

Flatting

The act of using abrasive paper to rub and smooth a panel or painted surface.

Flow (or Flow Out)

The ability of a liquid paint film to spread out evenly after application to produce a surface free from application irregularities.

Fog Coat

A paint coat applied at higher than normal air pressure.

Fluid Cup

The container on a spray gun which holds the paint. Can be mounted above the gun (gravity fed gun) or below (suction fed).

G

Gloss

The shine or reflection from a painted surface.

Ground Coat

Also known as undercoat for brushing Synthetics.

Also known as a background coat.

A special colour undercoat which is used with Pearlcoat finishes.

Guide Coat

A thin layer of paint of a contrasting colour used prior to rubbing down a panel. The thin surface of paint should be removed entirely by the rubbing down process, thus ensuring that the whole panel has been prepared properly and has no high or low spots.

H

Hard Dry

When the drying has reached such a stage that if desired, a further coat can be satisfactorily applied after flatting, OR, the paint film is hard enough to allow certain other operations to be carried out. For example the re-fitting of trim, or moving the vehicle out of the spraying booth.

Hardeners (or Catalysts)

Hardener or catalyst is the property added to the paint to set the paint film off very hard. It can be stoved or air dried. This product contains Isocyanate.

High Build Primer

A primer with high build properties, mainly used for covering filler.

Holding tank

A steel tank which is used as a reservoir for compressed air. The larger the volume of the holding tank the longer a given spray gun will work without the compressor cutting in again and compressing more air. Also known as an air receiver.

Humidity

The percentage of water in the atmosphere. Affects the speed at which the paint will dry. Can determine which thinners to use when mixing the paint.

I

Isocyanate

Hazardous fumes given off when spraying or using Two-Pack paints.

See the First Aid information at the front of this book.

See Two-Pack Paints.

L

Lacquer

A paint finish made up of pigment, resin and solvent, which dries quickly due to evaporation of the solvent.

Linnish

Term used to describe a sanding or grinding process when preparing a surface prior to painting.

M

Mapping

Also known as Ringing

The shrivelling of an edge of a repaired area, so that an outline of the repair, which has sunk, shows through the top coat. Caused by solvents attacking the edge of the repair.

Masking

Areas which are NOT to be painted are covered with paper (usually brown paper) or special sticky tape. This covering process is called masking. Hence masking paper (brown paper) and masking tape. It requires considerable time and effort to mask properly.

Matt Finish

A finish with no gloss.

Metallic

A paint containing metallic powder or particles.

Mist Coat

See also Guide Coat.

A mist coat can be applied when spraying Synthetics. It is the first colour coat misted on to give good adhesion.

O

Opacity

The ability of a coat of paint to obscure (or cover) an underlying surface.

Orange peel

A common problem which occurs when the wet paint does not flow properly on the panel after spraying.

Overlap

The amount by which each subsequent pass of the spray covers the previous pass.

P

PSI

Pounds Per Square Inch, the pre-metric measure of pressure, still in everyday use in industry. The metric equivalent is Kilonewtons per square centimetre. Also measured in Bars, where one Bar (or Barometer) is equal to 14.2 pounds per square inch.

Paint

A mixture of pigment, binder and solvent used to decorate and protect metalwork.

Paper Grade

A measure of the roughness of the abrasive paper. For example a grade of 80 would be very coarse, while a 600 or 800 grade of paper would be very fine. Grades range from 40 (production paper) or 80 wet or dry, up to 1500 superfine grade.

Pickling

Also known as Wrinkling, Puckering or Shrivelling.

Fault condition where the surface of the paint wrinkles like the surface of a prune.

Plasticizer

Plasticizer is a substance added to paint when spraying or coating plastics. This enables the paint to become more flexible and move with the plastic. This additive stops the paint drying rigid and avoids cracking.

Polishing

The action of rubbing a painted finish to a very good gloss (or shiny) finish.

Primer

A type of paint which has little colour content but which helps the bonding process between metal and colour coats.

Primer/Surfacer

A heavily pigmented primer used to fill in sanding marks and small surface imperfections.

R

Recoat

The action of going over a surface already coated.

Reducer

Solvents used to thin enamel paints.

Retarder

A slow drying solvent used to slow down evaporation and hence slow the drying of a paint coat.

Rubbing Down

See Abrasive, and Wet and Dry.

S

Satin Finish

A semi-gloss finish.

Scotchbrite

A fine wire wool used to prepare a surface for painting. Available in several grades such as fine and superfine.

Sealer or Isolator

A coating needed to cover synthetic paint to allow a recoat with any other material except synthetic.

Shrinking

The process where a newly sprayed paint contracts as it dries. See Sinkage.

Silicone

Ingredient in polish which is the greatest enemy of the spray painter. It prevents good adhesion when spraying.

Sinkage

The process where the paint sinks into a porous surface. Gloss is reduced by this process. It also shows imperfections such as stopper edges, scratch marks etc.

Solvent

Any material used to thin paint prior to application, such as thinners.

Spray-gun

A spray-gun is a device used to apply paint to a prepared surface. Professional spray equipment requires an air compressor to activate the spray-gun.

Stopper

A body putty used to fill defects such as

pinholes after filler has been used.

Striping

This is the most common problem encountered when spraying metallics. Generally caused by one of three things: insufficient overlap between coats leading to a "dry edge", or using cheap thinners which flash-off at the wrong speed for the paint, or poor gun technique.

Stoving

The action of drying paints by means of heat, rather than by natural air-drying. Stoving is often carried out in a specially heated stoving oven or spray-booth.

Surfacer

A coating applied after the primer. Used to build up the surface and fill minor scratches. See Primer/Surfacer.

T

Tack Coat

First coat when applying enamel. It is allowed to dry until sticky.

Tack Rag

A special cloth used to remove dust and other contaminants from a panel immediately before it is sprayed. The rag is impregnated with a sticky non-drying varnish.

Thinners

The solvent used to dilute many types of paint for spraying.

Thixotropic Agent

The thick material that settles at the bottom of the paint tin. This material must be thoroughly stirred to add body to the paint.

Top Coat

See Finish Coat.

Two-Pack Paints

A paint or lacquer supplied in two parts which must be mixed together in the correct proportions before use. The mixture will then remain useable for a limited period only.

See Isocyanate.

V

Viscosity Cup

A measuring cup used to measure the degree of thickness or thinness of liquid paint.

Wet and Dry Paper

A slightly misleading name given to a waterproof abrasive paper which is almost always used WET. Mostly used for rubbing down paintwork as part of preparation. More commonly used with a bucket of water to which some soap or washing up liquid has been added. The soap or washing up liquid helps to prevent the paper from clogging with removed paint particles.

About Paint

During the course of writing a previous book, ***"Restoring Small Fords,"*** I bought a 1965 Ford Cortina MK1, G.T. model. Much of the restoration work on that car was described and illustrated in the above book. Early in 1992 I began to ask around for estimates to have the car professionally resprayed.

The results were amazing. Prices tended to be around £800 for a complete, professionally finished respray. This included the fact that the bodyshell would be presented to the painter as a fully stripped-out shell with little finishing work to be done. A complete change of colour was also asked for when seeking the estimates.

It soon became apparent that I was not going to hand over £800 for a respray when I realised that a full set of professional quality paint spraying equipment could be assembled for around half that price. I also have three other cars in various stages of renovation so the decision to learn how to spray paint was made.

My labour is cheap when compared to a professional who has to earn his living. So if eighty hours of preparation time would be needed on a bodyshell then that would cost me nothing but eighty hours of hard work.

Once the decision was made I telephoned around for a few catalogues containing spray equipment. I rapidly began to get baffled by PSI, Bars, pressure drop, flashing-off, bleeding your compressor and so on. In other words there is a whole new vocabulary to be learned when taking up spray painting. We have assembled a Glossary of Painting Terms which begins on page 9. You need to read these terms and have an idea what they mean, otherwise you are quickly going to get lost in the jargon of spray painting.

Why Paint?

Why do we actually paint a vehicle? The three main reasons are:

1) Protection of the vehicle from corrosion (we can't all afford a stainless steel De Lorean!),

2) To present an attractive appearance (this could also mean a "custom" finish),

3) To use colour as a means of identification or advertisement for the vehicles's business use. Good examples of this which we see every day might be National Express buses, or the Post Office's Parcel Force vans.

How do we paint?

There are several methods available to apply paint. These are:

Brushing or rolling (commercial use), or Spraying (which includes cold spraying, hot spraying, airless spraying, and electrostatic spraying).

This book will tend to concentrate on cold spray painting.

Who can spray?

Well, almost anyone can spray paint, but he must spend a long, long time doing the preparation. Success is in the preparation, not in the spraying. That is why there's not one but TWO chapters on preparation in this book.

The only people who may NOT be able to spray are those suffering from chest problems such as asthma. Seek doctor's advice if you are in any doubt before investing in expensive equipment you may not be able to use.

What Equipment do I need?

The next thing I was told was to buy the best quality spray-gun I could afford. We are back to a balancing act between time and money. If you have unlimited time to spend either preparing the surface to be painted, and in polishing it afterwards, then you could buy the cheapest possible spray-gun and accept the extra work required. More realistically, if you want a really good gloss straight from the gun without spending days polish-

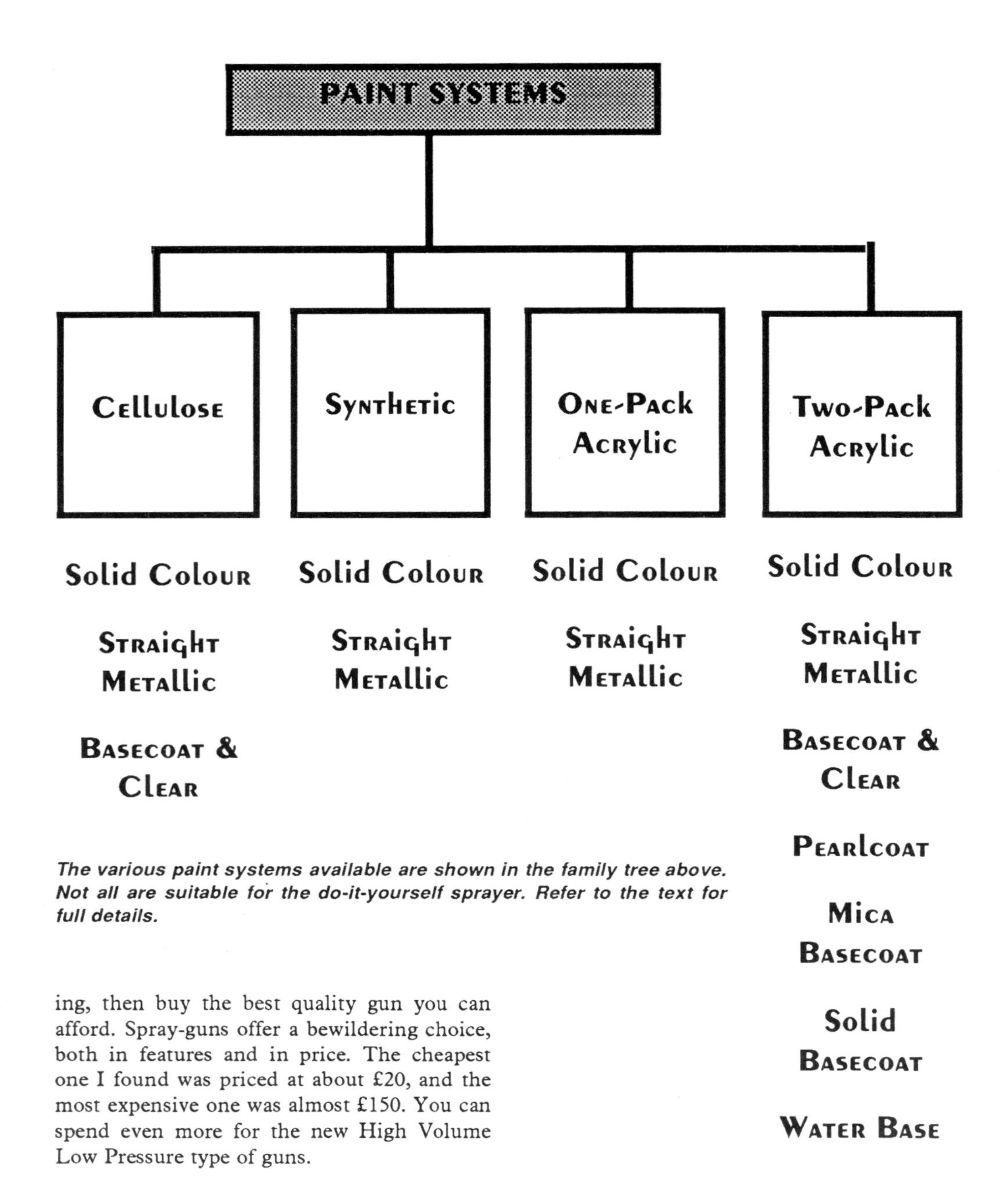

The various paint systems available are shown in the family tree above. Not all are suitable for the do-it-yourself sprayer. Refer to the text for full details.

ing, then buy the best quality gun you can afford. Spray-guns offer a bewildering choice, both in features and in price. The cheapest one I found was priced at about £20, and the most expensive one was almost £150. You can spend even more for the new High Volume Low Pressure type of guns.

What is Paint?

If I were to ask you the question, "What is paint?" I would expect several different sorts of reply, ranging from "dunno, mate!" to: "A chemical mixture used to protect and decorate a metal or other surface".

The second answer is of course much closer to the truth. Before we go any further in trying to learn to spray paint, we need to pause and look at what paint actually is.

Paint Ingredients

Automotive paints may vary in their properties and uses but they all have three components in common. These are Pigment, Binder (or vehicle) and Solvent (or thinners).

These are now explained as follows:

Pigment

Pigments used in the manufacture of paint are finely ground powders. These may be

Health and Safety

There are a large number of Health and Safety Executive publications covering motor vehicle repair, electrical safety, compressed air equipment, cutting and welding, body filling and preparation, and vehicle painting. These publications are available from HMSO Bookshops.

These publications provide guidance under the Health and Safety at Work Act 1974, and are generally intended for "professional" workshops. The amateur, working on his own is generally not required to conform to the above Act. However, most of the information provided is in a suitable format for the amateur to read and comply with. Most of it is common sense.

The following information is provided for information only.

Storing Paint

1) Paint and solvent can be readily ignited, and are usually toxic.

2) Keep a minimum amount of paint and solvents for the job in hand. The recommended maximum is 50 litres. These should be stored in a metal cabinet.

3) Keep lids on cans to stop vapour escaping.

4) Keep all paints and solvents away from sources of ignition such as open flames, electrical heaters, and gas flames. DO NOT SMOKE.

5) Ensure adequate ventilation.

ISOCYANATE PAINT

1) Vapours from Isocyanate paints (commonly known in the UK as "Two-Pack paints") are an irritant to the eyes and the respiratory system. They may cause asthma. Anyone suffering from asthma should NOT work with Two-Pack paints.

2) Spray Two-Pack ONLY in dedicated, mechanically ventilated spray booths. Do NOT use Two-Pack paint in the home workshop.

Symptoms of exposure to Two-Pack vapours are:

Sore eyes and running nose, sore throat, coughing, tightness in the chest wheezing and breathlessness.

Fatalities are rare but symptoms may occur up to 12 hours after exposure. Symptoms may clear up after a few hours but will return again if there is continued exposure. Consult a doctor.

Miscellaneous

For those wishing to construct their own spray booth, refer to HS (G) 67 Health and Safety in Motor Vehicle Repair. This publication provides basic information about the construction of a "home made" spray booth.

Ventilation is particularly important and professional advice should be sought if you are considering your own spray booth.

derived from naturally occurring minerals or they may be synthetic dyestuffs. Their properties are very important because they give the paint its covering power (opacity) and colour, and help to determine its durability. The actual pigmentation of paint depends on its function. Primer and fillers are chosen to give good build and easy flatting. In finishes (top coats) they are selected to give a lasting decorative effect.

Binder (or Vehicle)

This gives the paint film-forming properties, binding the particles of pigments together and providing adhesion to the substrates (the primer or surface being sprayed).

Thinners or Solvent

This makes the pigment/binder fluid mixture workable during paint manufacture. It also reduces the paint to the correct consistency for application by spray gun, brush or knife. The solvent mix is volatile, that is, it evaporates once the paint has been applied, leaving the pigment and binder to form the hardened paint film. Proprietary blends of solvents are used to reduce paints to application viscosity. These solvents are usually known as thinners.

Types of Paint

The basic paint types available on the market today are:

* **Cellulose,**
* **One-Pack Acrylic,**
* **Basecoat and Clear,**
* **Pearlcoat,**
* **Mica Basecoat,**
* **Synthetic,**
* **Two-Pack Acrylic,**
* **Water based paints.**

If we look at these in turn we should be able to discover their advantages and disadvantages. We also need to discover which are practical for the do-it-yourself sprayer to use.

Cellulose

Advantages

1) Can be used on anything from a spot repair to a full respray,

2) Fast Drying,

3) Easy to polish,

4) Can be used in any type of workshop or garage.

Disadvantages

1) Low filling properties, needs extra coats to achieve "build",

2) Strong thinners used could affect old paintwork,

3) Loses gloss after a time.

Cellulose paint has been used for many years and is the easiest for the amateur to use. It dries quickly and can be built up with many layers to achieve the finish required. Its main disadvantage is that it requires a lot of polishing to get the best finish. By a "lot of polishing" we are talking about perhaps four or more hours for the average car. This is one reason why Cellulose is no longer used by professional spray painters. A better finish can be achieved quicker by using newer materials.

Cellulose is also in danger from the environmental lobby. It relies on thinners to carry the paint from the gun to the panel being sprayed. This is a wasteful process and most of the thinners is required to evaporate or "flash-off" quickly into the atmosphere. Sooner or later the spraying of Cellulose is going to be restricted, controlled, or banned altogether.

Cellulose paint dries from the bottom outwards. This means that when you spray some onto a panel the paint next to the panel dries first, with the top coat drying last.

Cellulose paint can dry in less than 10 minutes depending on workshop conditions, at which point another coat can be sprayed on. If you are working in a professional spray-booth Cellulose can dry in anything from 1 to 2 minutes up to 10 minutes, depending on the temperature and the extraction system in use.

If you are working in an open workshop,

such as in a detached brick garage where an amateur may be working, you should be able to apply a second coat of Cellulose in around 5 minutes unless temperatures are very low in which case allow up to 10 minutes. You need to apply between 3 and 5 coats to get a good film thickness to polish on.

One-Pack Acrylic

This is similar to Cellulose but retains its gloss better.

Advantages

1) Similar product to Cellulose. Suitable for spot repairs or full respray.

2) Slightly quicker drying than Cellulose.

3) Easy to polish.

Disadvantages

1) High paint usage is necessary to build required film thickness.

Basecoat and Clear

This system consists of spraying on a colour or metallic coat (the basecoat) followed by a protective layer of clear material.

Advantages

1) Increased durability of paint finish,

2) Improved appearance,

3) High gloss,

4) Easy to fade out and lacquer a panel,

Disadvantages

1) More (and detailed) preparation required,

2) Extra time needed,

3) Polishing may be needed,

Pearlcoat

Pearlcoat paints represent the pearl finish you get in a pearl necklace. They are a three stage system where you put a white base on the background, followed by the actual pearl which is transparent and consists of aluminium particles, followed by a clear lacquer.

With this paint system you can only colour match edge to edge. You cannot fade out with it because the aluminium particles don't cover. The way of matching the paint is to identify the actual base white that you need. All base whites are different.

Outer layer of paint dry - inner layers drying

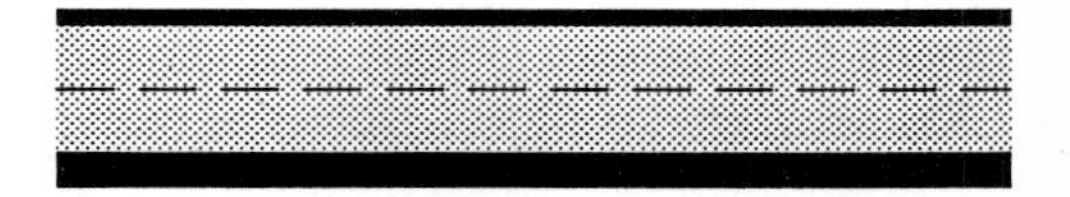

Body Panel

Outer layers of paint drying - inner layer dry

Body Panel

Synthetic paints dry from the outer surface inwards. The paint next to the panel dries last. See upper sketch.

Cellulose paints dry from the bottom outwards. The outside surface is the last to dry. This is illustrated in the lower sketch.

Mica Base Coats

Take a base coat colour and add pearlescent tints into the base coat and call it a mica base coat.

With mica you can fade out repairs. Trevor prefers mica basecoat to pearlcoat as pearlcoat tends to look -- from a distance -- as if the car hasn't been washed! With a mica basecoat it gives you a nice depth of gloss and seems to be a deeper basecoat gloss than the original Two-Pack basecoats.

The motor trade are tending to bring in lacquer more and more so you will find straight metallics or "solid metallics" as they are called will be replaced eventually with basecoat and clears.

The clear system is even coming in on solid basecoats so instead of spraying a normal red, some paint manufacturers are now bringing in what they call a solid basecoat so the solid colour you spray on tends to go to a mattish

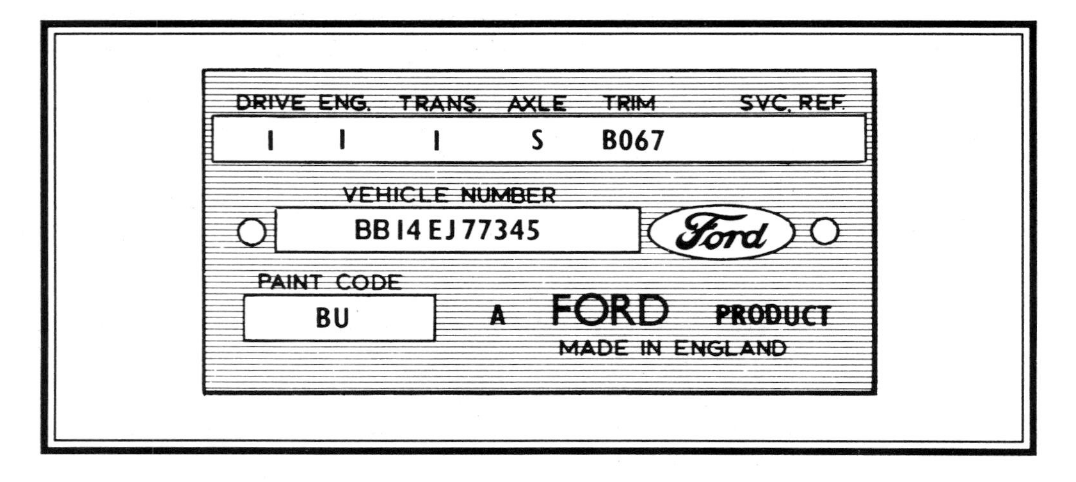

An ID plate from a 1960s British Ford. Together with all the mechanical data is information about the paint colour used on the bodyshell. Your paint supplier will have information about the colour and can mix the paint for you accordingly. Refer to data in the vehicle's Parts List for the name of the colour (for example Monza Red, or Ambassador Blue). Where two paint codes are given, one will be the roof colour on a two-tone vehicle. See text, opposite.

finish and you lacquer over the top.

Water Based Paints

Because of the solvents lost to the atmosphere when spraying paints (particularly Cellulose), the paint industry has devised water based paints. These have been available since the early 1980s and they considerably reduce the amount of solvent emission.

Advantages

1) Extremely easy to use, no extra training required,

2) Reduced solvent emissions, which will satisfy the environmental lobby,

3) Provides excellent fade-out,

4) Is non-flammable,

5) Has less overspray,

6) Has no unpleasant odours,

7) Needs less time cleaning guns,

Disadvantage

1) Intended for professional bodyshops,

2) Intended that the paint is dried by an air movement system.

Ask for information at your paint supplier.

Synthetic

Advantages

1) Used mainly for resprays or recoating of synthetic materials,

2) Very good level of gloss straight from the gun,

3) Does not react with any other product, (when applied OVER another product),

4) Very good filling properties,

5) Retains gloss level.

Disadvantages

1) Is an oil-based product so takes longer to dry,

2) Stays wet longer so causing recoat problems.

Cheap and cheerful and used for the proverbial "blow-overs" by back street garages. One major problem with this type of paint is that you cannot spray other types of paint on top of it. So if you have bought a vehicle which has been sprayed with synthetic paint, you will probably have to strip back to bare metal or apply some sort of isolator to prevent the two types of paint coming into contact.

Synthetic paints dry from the top inwards. This means that the outside or top surface of

How to find the Paint Code

On each and every car there should be an identification plate, (ID plate) giving information about the vehicle. See illustration, opposite. This includes details of the trim used in the vehicle and also paint information.

Refer to your vehicle handbook, or workshop manual for the location of this identification plate. Generally, it is located on the inside of the inner wing or on the bulkhead. If you can't find it there, check the documentation.

Having found the ID plate, what does it mean? Various headings indicate the type of engine, transmission fitted and so on, but you want the PAINT CODE and perhaps the TRIM CODE.

PAINT CODE. This describes the colour of the paint used on the bodyshell, for example on a Ford from the 1960s: A = Savoy Black, M = Ambassador Blue, AN = Monza Red, and so on. So when you order paint to have your car re-sprayed you ask for (for example) Ford paint code AN, Monza Red. Any worthwhile paint supplier will recognise the reference code and can mix the paint for you accordingly. For a full list of paint colours for your car refer to the relevant workshop manual and/or parts book.

The paint code may consist of a single letter or two letters. If you have a two-tone colour scheme, you may find two groups of letters. One will refer to the main colour, the other to the roof colour, for example. It is usually quite easy to work out which is which!

Remember a couple of things about this paint code. If you respray your BA (Ermine White) Ford in AN (Monza Red), then you keep the ID plate which says BA.

So if you are looking over a vehicle with a view to purchase, keep you eyes open for this sort of mismatch between paint code and actual colour scheme. In the UK if the vehicle has its colour changed this information should be changed on the Registration Document V5. If this has not been done, try to find out why. Has the vehicle been stolen, resprayed and now being offered to you? "Let the buyer Beware!"

A few paragraphs ago we mentioned TRIM CODE. This provides the colour and/ or combinations of colours used on seats, trim panels and carpets. There were probably hundreds of trim combinations available for each vehicle. This code is very complex and could fill many pages. Either write to the manufacturer for details of your car, OR look in the Vehicle Parts List.

The Owner's Club will undoubtedly have this sort of information available too. If you are changing the colour of the vehicle, you may want to consider changing all the trim to match the new colour. This could be difficult! While mechanical spares are thrown into a garage and left "in case they are needed" unwanted trim panels are simply thrown away. You will find the search for vehicle trim a very difficult one. The best hope is look through owner's club magazines looking for someone who is breaking up a car with the required trim.

the paint dries first and the bottom layer, next to the metal dries last. This can present difficulties in judging when a layer of paint is dry. A second coat can be applied almost immediately but needs to flash off for about 10 to 15 minutes depending on heat and extraction. The procedure to follow when spraying synthetics is to put a mist coat on, leave it to flash off for 5 minutes, followed by a full coat. Leave for about 10 to 15 minutes, then a second full coat can be applied. Then leave the job to dry thoroughly. If something goes wrong then a coat of synthetic must be left for about 24 hours before it can be rubbed down and re-painted.

In contrast, Cellulose may be ready for another coat in 2 to 10 minutes.

Two-Pack Acrylic

It may surprise you to know that Two-Pack paints have been on the market for more than

10 years. The name is actually misleading because it consists of three packs of material. These are the paint, hardener and thinner.

It provides an excellent gloss straight from the gun but the operator requires an air-fed mask to protect himself from Isocyanate which can cause serious damage to your health.

Two-Pack paints are almost always described as being unsuitable for the amateur to use. However, in Chapter 11 advice is given for their use.

The decision to provide this information was taken on the basis that it might save you from some damage which might occur if you are daft enough to try to use it without taking the proper precautions.

However, you would be most unwise to consider using Two-Pack paints without access to a professional spraying booth with suitable extraction equipment.

Drying Methods

I bet you never thought there was much to discuss about paint drying! Well there are four stages in the drying process, and these are discussed in the next few paragraphs. The amateur spray painter will need to be aware of these stages to prevent disaster overtaking his spray job.

The transition of paint film from a fluid to a solid state is caused by physical and/or chemical means. For example, air drying, heat drying (such as oven drying, baking or stoving) and radiation drying are all methods used to dry paint. The drying stages are:

1) Dust free -- when dust no longer adheres to the painting surface,

2) Surface dry -- when the paint is dry on the surface but is soft and tacky underneath,

3) Tack-free -- free from stickiness even under pressure,

4) Touch dry -- when a very slight pressure with the fingers does not leave a mark or reveal stickiness.

Fast Drying

Fast drying is a common phrase used in spraying. However it does NOT mean using cold compressed air. This technique is not commonly used but you can flash-off a light coat with air from the spray-gun if you want to apply another one straight away. Use of other artificial sources of heat, such as a hair drier, are not recommended.

Fast or forced drying can be done but only by using very special infra-red heaters.

The problem with force drying is that if not done correctly the top coat tends to crust off while the underneath coats are still drying out and tending to move. This gives you a false sense of security. If the top coat crusts over it is hard to flat and recoat with another material or re-spray as the underneath coats still have to dry out. The result is that you get cracks.

Fast drying techniques are NOT recommended for the amateur.

What Colour?

There are many old wives' tales about which colours are easier to spray. If you are a qualified sprayer then no colour is easier or harder than any other colour.

The only colour which might be described as difficult is yellow, because of coverage. Yellow has poor opacity. It might need two or three coats of yellow to achieve the same coverage as a black or red colour which might just need one coat.

Whites are difficult as well as they reflect the light and it is often difficult to see where you have sprayed.

There is a fallacy about black being hard to spray. Black is very easy to spray but the reason people stay away from black is because of the preparation. Preparation must be 110% as black shows up every little mark. (Tommy can personally confirm that this statement is true!)

So if you want to skimp on your preparation, don't consider using black. Generally a dark colour or high gloss (or both), reflect any defects in preparation.

White on the other hand tends to hide any repairs. So if you have a lots of repairs use white!

Having said all that, every sprayer has a "bogey" colour which gives him more problems than others. For example, yellow again! People try to cover in one coat as they do with red, black or blue. In a dark colour the pigment is a lot stronger and will cover in one or two coats. With yellows the pigment is very weak and this leads to the need for three, four or five coats to cover.

If you "attack" a coat of yellow to try to get coverage in one coat, you may run into problems with runs as you apply too much material.

Safety

Before we move on and look at the equipment needed to spray paint, it is worth having a look at what safety measures may have to be taken.

Health and Safety is discussed on page 17, but for now have a look round your garden.

For example, if you are going to spray in your garage is there adequate ventilation? Do you need to install an extractor fan?

That fish pond over in the neighbour's garden... are you going to kill his fish with overspray from your gun?

Pets need to be considered. Are you going to have puppies or kittens running around your feet when you are trying to spray? A young animal may be put off by the noise and fumes, but on the other hand may decide that your just-resprayed bonnet would make a nice place for a sleep.

Young children too need to be kept out of the way. While there is no immediate danger from paint fumes, it is not recommended that anyone stand inhaling them for any length of time. That is why you will be wearing a mask. Ensure that children are kept out of the way when spraying.

All paint materials including thinners, are potential fire hazards.

Paint fires burn ferociously. Vapour from thinners can be ignited easily by a careless cigarette, electrical spark (such as from an electric motor), or an oxyacetylene welding torch.

DO:

Provide good ventilation,

Keep electrical appliances away from thinners or freshly sprayed paint,

Keep spraying areas clean and free from inflammable waste such as old masking papers, rags etc.

Keep a fire extinguisher handy.

Dispose of waste materials safely. Rags may be impregnated with solvent, masking paper also could be highly inflammable due to paint or solvent absorption.

DON'T:

Smoke cigarettes when spraying,

Allow onlookers to smoke,

Use electric extractor fans, unless they are specially built for use in a hazardous area.

Don't switch electrical equipment on or off when spraying or during the flash-off time. Sparks from electrical switches could ignite thinners.

BE SAFE -- NOT SORRY!

How much does it cost?

At 1993 prices Cellulose ranges from about £10 per litre for cheaper material, to £12 to £14 per litre for reasonable quality paint, and up to £18 to £20 per litre for absolute top quality material from one of the big manufacturers.

Two-Pack is going to cost around £30 per litre. Remember that Two-Pack will give better coverage than Cellulose so you should need less material.

Paint is expensive, so remember to shop around and see if you can negotiate a discount from your supplier.

You probably need about 3 to 4 litres of Cellulose to respray an average saloon car. If you are doing a complete respray, including door posts, and the insides of boot and bonnet then reckon on 5 litres or more.

If you are spraying a complete vehicle, we recommend that you work out how much paint will be required, allow a bit extra for the "just in case" factor, and buy sufficient paint in one batch from the paint supplier.

Tools & Equipment

The equipment needed to spray-paint your car, van, or truck can be described in a very short list. An air compressor is needed, plus a spray-gun, a quantity of paint and thinners, plus a list of sundries. These sundries are listed and described later in this Chapter.

You also need somewhere to spray.

Health & Safety

If you are buying a compressor for PERSONAL USE ONLY, in other words you will never spray someone's car for money, then you are almost exempt from Health and Safety rules. Your neighbour may still complain about noise, or overspray but from the Health and Safety point of view you have no further problems.

However, if you intend to set up a workshop at home to spray cars on a commercial basis, you are faced with many rules and regulations. Seek professional guidance.

Spraying Equipment

As with any technique new to you, there seems at first sight to be a bewildering array of bright, shiny equipment available. How much of it is really necessary, and how much is just nice to have?

The single most important items you need are a compressor and a spray-gun. If you only want to spray one car, then consider hiring the equipment. Look around the Yellow Pages for Hire Shops and phone round for some prices. You will need to hire the equipment over a period of several weeks, since you cannot hope to do a respray in one sitting.

For those who are new to spraying but will do more than one car, buying the equipment will be the best choice.

Specialist businesses such as Machine Mart have a wide range of compressors and equipment and are always happy to talk to customers on the phone prior to a purchase.

Compressors

Compressors come in many shapes and sizes, so it is quite understandable if you feel bewildered by it all. Basically the compressor squeezes the air which is then fed through a hose and makes the spray-gun spray the paint onto the car.

The type of gun you buy and the length of air hose you will be using, determines the size and power of the air compressor required.

For example, the top of the range De Vilbiss spray-gun, the JGA 30EX requires 10.6 cubic feet per minute of air at 50 pounds per square inch.

If you try to run this gun off a small compressor which cannot supply this amount of air, then the gun will spray properly for a moment or two then peter out as the air supply dwindles. If you shut off the gun and allow the compressor to catch up, the gun will work properly again for a time before the process is repeated. This will give rise to a very poor spray finish.

Since you are aiming to buy the best possible spray-gun you can afford, you must buy an air compressor which is a suitable match for the spray-gun. Take advice from your supplier.

A smaller, cheaper spray-gun may only require between 3 and 6 cubic feet per minute. This would allow you to use a smaller compressor.

Buying a larger, more powerful compressor is great for a top quality spray-gun and hence a top quality finish, but it creates another set of problems.

If you buy a compressor with a 3 horsepower motor to drive the required compressor, do you need any special electrical wiring installed? Find out before you buy. Industrial compressors use what is known as three-phase electricity, and this is not normally

This portable compressor is available from Machine Mart in the United Kingdom. This machine offers a maximum working pressure of 150 psi, and a 50 litre holding tank. It does not need any special electrical supplies and just plugs into normal mains electricity.

available in a domestic installation.

If you want to eliminate all electrical problems you may wish to consider a petrol or diesel-driven compressor. These must be heard running before purchase, in case there is a noise problem.

Petrol or diesel generators may also cause you problems because the exhaust gasses must be kept away from the compressor air intake, as you do not want contaminated air being fed to your spray-gun.

Compressors may be either portable or fixed, and again, you will need to sort out some accommodation for your compressor if it is going to be fixed equipment. Noise, and air filtration are the first two problems to be sorted out, as you do not want the compressor in the same workshop as you are, swallowing all the overspray and compressing it and passing it down through the gun again. The compressor needs a good, clean source of air supply.

When choosing a compressor, check the following points regarding maintenance. If you choose a compressor which sits on wheels, it is regarded as portable and if a problem develops while under guarantee you must return it to the shop. If on the other hand you buy a similar compressor which does not have wheels it is regarded as either "industrial" or "static" and a technician will come to your premises to carry out repairs. This point was not made clear when I purchased my compressor.

Remember to check the oil in the compressor regularly.

Holding Tanks

Holding tanks, or air receivers store the compressed air until it is required. Ensure that your tank is kept in good condition.

The size of the holding tank should be in proportion to the size of the compressor. However you are spared the worry of working out the correct combination by buying an already assembled compressor and holding

This gauge shows the pressure of air stored in the compressor's holding tank. In this case it is set to 150 psi. There is a cut-out switch on the compressor's motor to stop pumping when 150 psi is reached. When pressure drops by about 20 psi the pump switches back on again.

tank. That's the sensible thing to do.

The holding tank performs several functions. First it helps to cool the air, which has become heated due to the compression process. Secondly, it helps to smooth out any peaks and troughs in the air supply, by acting as a reservoir for the compressed air.

For example, my compressor will store 50 litres of air at a pressure of 150 psi in the holding tank. When this pressure is reached, the electric motor is cut off and no more air is compressed. As I use my spray-gun or other air powered tools, the air pressure in the tank will drop. When it gets to about 130 psi, a sensor detects the pressure drop and switches the motor on again. It then "tops up" the air tank. This process is repeated as long as you continue to use the air supply. Remember, when spraying I am only using about 60 psi, so there is adequate reserve in the tank.

Finally, remember to drain off the holding tank at the end of each spraying session or each evening after use. You will find some sort of drain screw arrangement on the bottom of the holding tank. Apart from releasing any air stored in the holding tank, you will drain any water which has collected in the tank. The water got there as the hot air from the compressor cooled down. It is a by-product of compressing the air.

Filters and Drains

When air is compressed it heats up. For those interested in the technicalities of this, refer to Boyle's Laws. The best analogy to this is your car engine which compresses the fuel/air mixture and heats it up to help combustion.

When this heated air is cooled, water forms and this is the enemy of the spray painter. For this reason you must ensure that your equipment has a drain fitted so that water can be drained off from the air holding tank. Professionals may use a water trap in the air line to prevent water being carried along with the compressed air.

Drain off the holding tank at least once per day when using the compressor.

Compressed Air Safety

The following rules apply to all users of air compressors:

1) Compressed air is dangerous. Do NOT direct compressed air at people or animals.

2) Do NOT operate any compressed air machinery with safety guards removed.

3) If a mechanical or electrical problem occurs, do not try to fix it yourself, contact your local dealer for professional advise.

4) If a problem develops, switch the equipment OFF and bleed off any air in the air tank.

5) Do not leave the air tank full of compressed air overnight or when moving the equipment. ALWAYS bleed off the compressed air each night or after a spraying session.

6) Ensure an adequate supply of clean, fresh air is available for the compressor. Check air filters at recommended intervals.

7) Ensure that electrical supply cables are in good condition and not damaged, cut or frayed.

8) Ensure that air hoses are in good condition and not damaged, cut or frayed.

9) Remember that the cylinder of the air compressor gets very hot in use. Do NOT touch the compressor when in use and allow a cooling down period before touching the equipment.

10) When spray painting remember: Never spray where a naked flame or source of heat could ignite the vapour, and always ensure that the spray area has adequate ventilation.

If you have a run of fixed piping you will need to have another drain at the end. You cannot be too careful about removing water from the compressed air supply.

As stated above, the compressor needs a good supply of fresh cool air. This supply needs to be filtered and care should be taken to keep the filter clean by following the manufacturer's recommendations.

Hoses and Couplings

Having sorted out your air supply, the next thing to be considered is how to get the air from the holding tank to the spray-gun.

Flexible hoses are normally used for this purpose, but in a commercial installation metal pipework may be fixed to the wall instead, with couplings sited at convenient points around the workshop.

Remember that compressed air loses pressure the longer the pipe run (another variable here is the inner diameter of the hose). Refer to the Tables in the Reference Chapter, at the end of the book, for pressure drops.

So, if you have set the pressure at the compressor to 60 psi, and are using a 5/16 inch internal diameter hose 25 feet long you will get 60 minus 19 psi, which is 41 psi at the end of the hose. If you need 60 psi at the spray-gun, set the pressure at the compressor to 79 psi. In this case 80 psi would be easier to set on the gauge.

The answer is to site the compressor close to the work area to ensure least pressure drop, while maintaining a good supply of clean fresh air to the compressor.

For general use a 10 metre length of hose with an internal diameter of 1/4 inch will be ideal for the amateur sprayer.

5/16" is better.* 3/8" better still!*

At the compressor end of the hose, make up a permanent coupling, as shown in the photographs.

At the working end use a quick-release coupling. This will prove invaluable when you want to disconnect the spray-gun for cleaning, changing colour etc. It also helps when you want to use different compressed-air power tools.

* Better so long as the compressor is a powerful one. When using a small compressor a 1/4" hose is better so long as the hose is fairly short. If a small compressor puts out 140 lpm and the gun uses up to 200 lpm, a 1/4" hose is essential to restrict the full output of the gun.

Chapter Two

Compressor Care

Check oil daily -- before using the compressor.

Drain holding tank after use. (Air being compressed generates heat. When the hot air cools, water vapour is given off and this collects in the holding tank.) Remember to drain water in Winter as ice could form in the holding tank if the equipment is kept in a non-heated garage or workshop.

Wipe down air lines after use. Remember that any dirt on an air line could find its way onto your spray job.

Fire Extinguisher

A dry powder fire extinguisher is recommended.

Do NOT use water on a paint fire as this may spread the fire to other areas. Contact the Fire Brigade immediately if a fire occurs.

Paint fires burn ferociously. You cannot be too careful.

Gun Types

Spray-guns fall into several distinct types. First of all there are guns which use compressed air to atomise the paint. Then there are electric-powered guns. Most professional equipment is powered by compressed air, and in this book we will focus on this type of equipment.

There are several different types of compressed air spray-guns available. Two common types are known as the suction feed, where the paint pot is on the bottom, and the gravity feed gun where the paint pot is on the top.

The suction type is generally preferred for car resprays, while the gravity gun is often used where smaller component parts are to be sprayed. It is also commonly used on production lines where the same items are sprayed as they pass along the line.

The more common type of gun used for vehicle re-finishing is the suction fed gun. This has the paint pot under the gun. The paint is lifted from the paint pot by blowing the compressed air over the top of a pipe which is inserted in the paint. The low pressure created by blowing air across the top of the tube lifts the paint into the air stream where it is atomised and deposited on the panel being sprayed.

The main disadvantage of this gun is that the operator has to hold the weight of the gun plus the paint in his hand. This can cause fatigue.

Paint spraying equipment can also consist of a lightweight gun which has a remote paint container. This, again, is often used in industry where conveyor belt spraying is carried out.

Spray-Guns

When I first started to look for a spray-gun I met with two quite distinctive (and opposite) pieces of advice. The first said, "Buy a cheap gun to start with as you won't know the difference until you have gathered a lot of experience."

The other voice said, "Buy the best gun you can afford. You will get a better finish from the gun and have less polishing work to do."

I was also told to buy one gun for spraying primer (a cheap gun) and an expensive gun for colour coats.

In the end I bought a top of the range De Vilbiss gun because I intend to spray at least two or three cars before I give up restoring old cars. I did not buy two guns.

When buying the spray-gun, check which "set up" has been fitted. This means, which air cap, air needle and fluid tip. What is fitted determines what type of paint can be sprayed.

The first time sprayer will probably want a set up for Cellulose, so air cap number 30 will be the best choice. Later, when you become more experienced and wish to try spray painting with a different type of paint, you will have to change the set up of your gun. See the Table in Chapter Five for details of

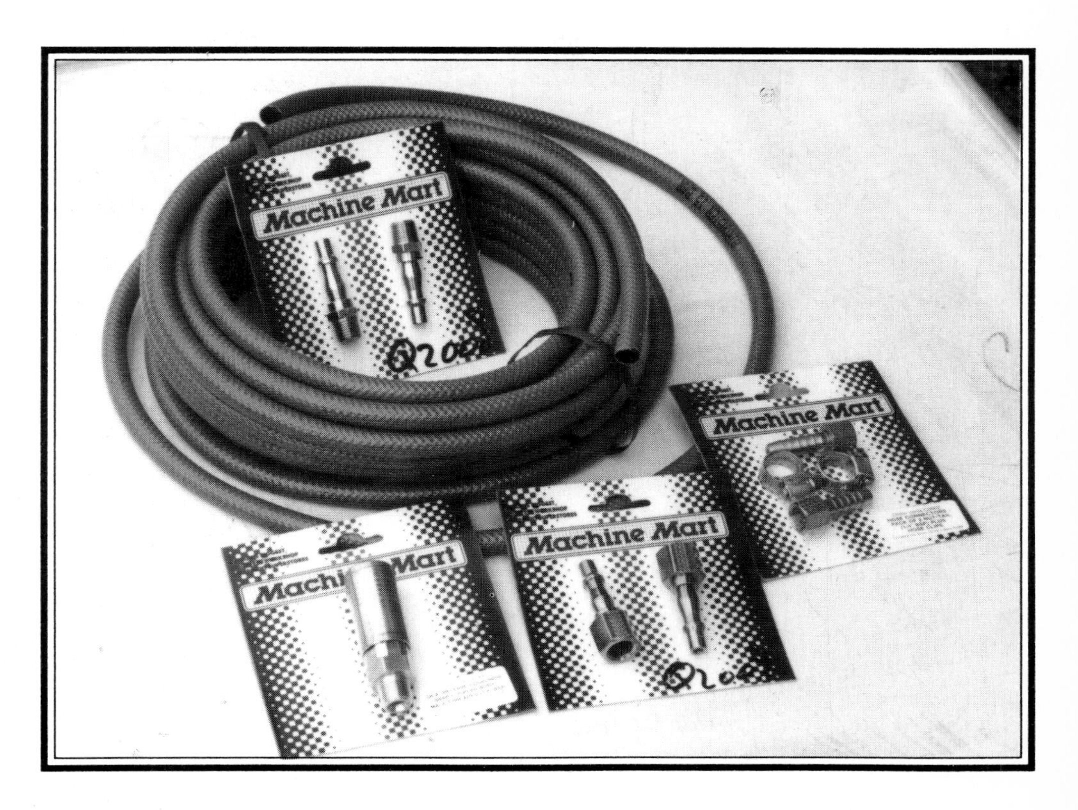

Ten metres of 8 mm air hose suitable for supplying spray-guns and other air-powered tools. The quick-release coupling (left) is fitted to the tool end of the air hose. Once fitted and secured it provides a quick and efficient method for you to uncouple the spray-gun from the air line when refilling or cleaning the gun.

spray-gun set ups.

Electric Spray-guns

So far we have concentrated on the professional set-up where compressed air is used to spray paint. There is another much cheaper alternative which may appeal to the amateur. That is the airless, or electric sprayer, very often known by the maker's name "Burgess."

These sprayers have been around for many years and many good jobs have been done with them. However they have many limitations and care should be taken not to get carried away with enthusiasm for them. I had one a few years ago and a friend used it to excellent effect on a Volkswagon respray. On another occasion I used my Burgess to spray a bonnet panel for a Ford Capri.

However, while they may be used with Cellulose paint, they are better suited to spraying materials such as varnish or Hammerite paint to protect chassis and sills.

If you choose to use one for a respray, ensure that the mixture of thinners to paint is exact. Get it wrong and the job will suffer badly.

Spray Booth

Where are you going to spray? If you have a brick garage with room to move all round the car, then this can be used quite successfully. If you only have enough room to get at one side of the car at a time, then you can still use this garage but will be forced to move the car when it is half sprayed. This could introduce several problems.

While the professionals use a custom-made spray booth, the amateur can with some success, copy this quite cheaply. Your local garden centre may have what are known as "poly-tunnels", which are simple plastic greenhouses consisting of several steel loops with their ends in the ground, and a large sheet of heavy clear polythene stretched over the hoops.

These can be used for painting in, but make sure you have enough room to walk round the car when choosing a tunnel. The main problem is that they do tend to get very hot inside in the summer. I have had temperatures as high as 95 degrees F in mine this year. This is too hot for painting, so I can only paint in the morning or in the evening when the temperature is lower. However, this is all I have available, so I have to adapt to it.

The last alternative is the driveway respray, where everything is done outdoors. In Africa, America and Australia this may be quite acceptable depending on the season, but in the UK you may have good spraying weather one day, followed by rain the next.

You can spray a car in the driveway, but it is liable to be an unhappy experience as you are completely at the mercy of the weather. Plan to do it this weekend, and you can be sure it will either be windy or be raining. The temptation is then to "get something done anyway," and the result is a shoddy job which needs considerable time to put right. My recommendation is to hire a shed or garage if you do not have one. The end result will be worth the extra expense.

Heating

Spray painting is best carried out at about 20 degrees Centigrade (approximately 68 degrees F). See the illustration on page 79.

You may need to invest in some form of heating for your garage. Remember the fire risk. You CANNOT risk using an open flame source of heat unless you switch it off and ensure that the heating source has cooled down prior to spraying. Thinners have a flashpoint as low as 23 degrees Centigrade, so you have to be extremely careful.

A hot water radiator would be ideal, but we do not all have the luxury of a large enough garage heated by hot water radiators.

If you consider heating by electricity, then you have similar problems. The spark from an electric motor could ignite your thinners. Again, think very carefully about this.

Special electric fans are available for spraying, but their motors are specially protected so that sparks are limited and sealed inside the motor. These motors are said to be "intrinsically safe" and of course cost a lot of money.

Some people may wish to use paraffin heaters to warm the workshop. Remember that paraffin burners have two serious drawbacks; firstly, they use a naked flame, the dangers of which have already been discussed. The second problem is that for every gallon of paraffin burnt, 1 pint of water is released into the atmosphere. This excess moisture could cause blooming or microblistering in your finished paint job.

It is essential to keep moisture out of air as much as possible. The best forms of heating are dry heaters such as infra-red heaters or propane gas.

Remember with a propane heater there will be a naked flame. This flame must not be allowed into contact with spray vapours otherwise an explosion could occur. Switch off all forms of workshop heating unless you know it is safe.

TURN THE HEATING OFF BEFORE SPRAYING.

Personal Protection

Under this heading comes breathing masks, overalls, head protection, cotton gloves, boiler suits and so on.

Breathing Masks

There are several types of masks available on the market. Look at the range available in the Machine Mart catalogue, or visit your nearest paint factor.

The three main types are the filter mask, the charcoal filter mask and the air-fed enclosed face mask. We will look at these in turn and point out their good and bad points.

Cotton Mask

Many a home respray has been done with the sprayer wearing a simple cotton dust mask. However, you are fooling yourself if you think they provide suitable protection. They are intended more for filler work, such as sanding down plastic body filler, than for

The rubber sanding block. You'll probably want two or more of these, especially if you get some friends round to help with the rubbing down! These blocks take 1/3 of a sheet of abrasive paper. Being rubber, they will last for ever and can be used wet or dry.

paint spraying. They cost just a few pounds, but are better left for the job they were intended for - keeping dust out of your throat.

NOT RECOMMENDED FOR PAINT SPRAYING!

Charcoal Respirator

Similar to the war-time gas mask, the charcoal respirator may have one or two filters. Each contains activated charcoal which filters out any airborne particles which would otherwise enter the sprayers lungs. These respirators cost from about seven or eight pounds for the single filter type to around eighteen pounds for the double. Periodically, new filters must be fitted and these can be fairly expensive. Check this when buying a respirator.

NOTE: These respirators are NOT suitable for use when spraying Two-Pack paints. An air-fed mask MUST be used when spraying Two-Pack paints.

Air-fed Masks

Air-fed masks are essential if spraying Two-Pack paints which give off Isocyanate. They consist of a full face mask, with eye protection complying to BS 2092, Grade 2, and provide breathable air from your compressor.

However, air from a compressor MUST be filtered before it can be breathed, so a filter system is built in. The equipment requires about 4 CFM from your compressor, so take this into account when calculating compressor requirements.

When you wear this equipment for the first time, you will be put off by the "hissing" noise of the air supply. Try to ignore this and concentrate on the spraying. Wearers of spectacles may find this air-fed mask useful as your glasses will not steam up when you start to spray!

An adjustable body file which can be used for flatting plastic filler, or filing metal. The turnbuckle arrangement allows the blade to be curved to suit the surface being filed. Various blades are available.

Head Protection

Imagine you are spraying the final top coat of your respray. You have spent perhaps 100 hours to get to this point. As you apply the final coat a hair drops from your head or beard onto the car. "Oh bother," will probably be your response!

You can overcome this problem by wearing a balaclava type helmet which covers all the hairy parts and keeps hair off paintwork. Remember that the helmet will also keep overspray off your hair.

Overalls

For less than five pounds you can buy a set of disposable, one-piece painting overalls which keep you protected and keeps bits of fluff off the paintwork.

Normal nylon overalls may be used, but they must be spotlessly clean and you may find that the disposable ones are superior for painting.

Cotton gloves

Once you arrive at the stage of having a "perfect panel" (see Chapter Four) you do NOT want to touch it with sweaty fingers. Even one single fingerprint can spoil the respray on a panel, and to put this right may need considerable effort. So, if a fly lands on your ready-to-spray-panel, and you wipe it off with a sweaty finger you could have just added many hours to your respray.

Better to invest a few pounds in a pair of cotton gloves, or even fine rubber gloves. When the panel is ready for final colour or top coat spraying, wear the gloves and don't let anyone else near the car.

Shoes

There are two schools of thought here. For Health and Safety reasons the professional sprayer will always wear steel toe-capped shoes or boots to protect his feet in the workshop.

However, the home sprayer is probably going to look for something comfortable, such as an old pair of trainers. You must make the decision as to what the risks are in your home workshop.

Both amateurs and professionals are agreed on one thing. Do NOT wear steel-studded boots, as a spark could ignite the thinners being used with fatal results.

Having now equipped the spray painter, what other tools or equipment are required to tackle the respray?

Sundries

This list could be endless, but I'll try to keep it short. The main items of equipment are for removing unwanted paint, and sanding down surfaces.

Wet and Dry Paper

You will need large quantities of wet and dry abrasive papers, and a selection of rubber mounting blocks. Refer to page 49 for details of abrasive papers.

Rubber Sanding Blocks

I use a 3M rubber sanding block which shouldn't ever wear out and is comfortable to hold. Little metal spikes hold the paper in place, and you can tear a sheet of abrasive paper into three pieces, each of which will fit the sanding block. Draper Tools also supply a similar block. Refer to the photographs.

"Two-by-Two" Sanding Block

Although you can buy professional quality sanding boards of all shapes and sizes, there is a cheaper alternative -- the piece of two-by-two timber lying in the corner of your garage. You need a piece about nine inches to 1 foot long depending on the job in hand. Wrap some wet and dry paper round the two-by-two and you have a cheap and adaptable sanding board.

Other variations of this could be larger or smaller pieces of timer, round dowelling sections etc. All can have wet and dry paper wrapped round them, and each size or shape is good for a particular application. Don't overlook readily available items like these.

Body Files

There are several types of body file available, but the one in the photograph is the type which is adjustable. You turn the turnbuckle in the middle and one way makes the surface of the file convex, and turn it the other way and it becomes concave. In this way you adjust the tool to closely match the contour of the panel being filed.

This type of file is very useful for removing excess plastic filler, giving a quick smooth action.

The other use of the body file is to file a series of marks in one direction across a panel, then another series of marks at right angles to the first. If you have any low spots on the panel they will show up by not being shiny. (That is being below the level of the rest of the panel, they will not be reached by the file). This gives a good indication of where more hammer and dolly work might be needed.

Many different blades are available for the body file. Ask at your tool store.

There are smaller and cheaper tools available which will quickly remove excess plastic filler. These tend to be a bit coarse in their approach, rather like a cheese grater, but in some situations they are exactly what is required. Have a look at what is available before you purchase this type of tool. The most expensive tool may not be necessary for the amateur where time is NOT money!

Wire Brushes

Plastic Body Filler can be removed with a rotating wire brush in an electric drill. I have used this technique before, it does work, but you are probably better off using a body file intended for the job. However, if you do not have a body file then the wire brush can be used.

The main use of wire brushes, either rotating or a hand brush, is in removing rust from metalwork.

They are quite cheap to buy and are quite effective. Keep brushes clear of underseal otherwise you will end up applying the remains of an underseal to an otherwise clean panel. Alternatively, keep a brush for use on underseal only.

Grinding disks for the 4" angle grinder. These are very coarse, being about 80 grit, and are used for grinding down welds. Various types and sizes are available but it will work out cheaper if you buy disks in bulk.

Plastic Body Filler

I have used David's P38 for many years and never had a problem. There are many other fillers available on the market so if in doubt take the advice of the supplier or motor factor. Use the brand leader and be guaranteed good results. There is nothing worse than all your filling work falling out as happened to me when I tried another -- more expensive -- type of filler.

There is a section on using plastic body filler in the next Chapter.

Lead Body Filler

Lead for car bodywork consists of 70% lead and 30% tin. The use of lead died out for a number of years as plastic fillers came onto the market. Many enthusiasts are now returning to lead. There is a measure of skill required which is satisfying when you have accomplished it. There is a section on using lead in the next Chapter.

Measuring Stick

A measuring stick is a graduated stick, very like a ruler, which allows various ratios of paint and thinner to be mixed accurately. Make sure you get the correct type for the kind of paint you will be using.

Viscosity Cup

A viscosity cup is a special cup with a hole in the bottom. When you mix paint and thinners at various mixing ratios, the fluid will flow through the hole at differnt rates. These flow rates are quoted on the manufacturer's Data Sheets.

Electric Drill

This is your old friend the electric power drill. It will drive a variety of attachments, including drills, grinding stones and several kinds of rotating wire brushes. It would be very difficult to do any serious work without access to a power drill, but again you might be able to buy one second hand. I use it mostly for drilling and wire brushing as I am lucky enough to have the next tool on the list, the angle grinder.

Angle Grinder

If ever a tool scared me half to death its the angle grinder! It needs constant care during use, as the grinding disc rotates at about 10,000 RPM which is FAST. It is a very effective tool for removing rust and grinding down welds. Depending on the grade (or coarseness) of the sanding discs you can get a very smooth finish very quickly.

WEAR GOGGLES and protective clothing when using it. Don't leave it lying around unattended, as a child could be seriously injured. It has quite a kick as it starts up. Bits fly about so fast, you can do permanent damage to yourself and the car before you realize what is happening.

Be very careful selecting attachments for this tool, as wire brushes have to be specially made to rotate at the high speed of this tool. I once used the wrong type of wire cup brush when removing rust from the underside of a car. The brush virtually disintegrated in about 5 minutes due to the high speed. This was bad enough, but for the next year little pieces of wire kept appearing in the car, in the driveway, and in my overalls!

I've also had a sanding disc break up on me, as it snagged on a bit of ragged metal. The disc fragmented and flew in all directions.

The smallest angle grinder (known as the 4 inch, because of the size of the discs used) can be bought from about 40 pounds.

Sanding discs vary in price depending on quality and coarseness. Typically, you might get 2 for £1. You can get these cheaper if you buy a box of twenty through a friend in the motor trade!

Let me end the discussion on the angle grinder with another warning. The little sparks given off when grinding are particularly nasty. ***DON'T LET THE SPARKS HIT ANY WINDSHIELD OR GLASS.*** The sparks embed themselves in glass, and you can imagine what this will do to a set of windscreen wiper rubbers. If you don't take care over this, you will cost yourself a lot of money.

The angle grinder is extremely useful for grinding down welds when a coarse disk is fitted. It can also be used to "linnish" a metal panel prior to applying body filler.

Verdict -- very useful, but potentially very dangerous! Wear goggles, gloves and eye protection when using this tool.

Air-Powered Sanders

There are many air powered sanders available, but check the prices before you buy. You can save quite a bit of money if you shop around for these sort of tools.

The advantage that air has over electricity is safety. Quite apart from the damaged 13 amp plug wrapped in sticky tape, or the oversize fuse in the plug, and the extension cable cut and exposed, there is still the risk that you will switch on an electrical appliance while thinners are still in the atmosphere. The spark from switching on can ignite the thinners with fatal results.

Air powered tools are much safer. The amateur may have suitable mains powered tools collected over the years, but these must be maintained in tip-top condition and NEVER used where they might cause paint fumes or thinners to be ignited.

Safety first.

Orbital Sander

Orbital sanders can be electric or air-powered. The professional is probably going to have a range of air-powered tools available, while the amateur will have some electric powered tools.

Orbital sanders are extremely useful for sanding large flat areas. The sandpaper is moved by the action of the motor, and this results in a smoother finish with less build up of dust to clog the paper.

I find them particularly good on doors, flat areas of wing and the large flat areas often found on front and rear valances. They can be used on curved surfaces, but care needs to be taken unless you want to gouge a groove in the material being sanded.

They generally hold one third of a sheet of sand paper, and are always used dry! Various

TOP. Co-author Tommy Sandham models the charcoal-type spray painting respirator. This type of respirator is essential to protect your lungs from paint and thinner fumes. This respirator is NOT suitable for use with Two-Pack paints.

BELOW. Tommy's De Vilbiss spray-gun, chosen after listening to lots of different advice.

The Black & Decker 1/3 sheet electric sander. This machine operates from mains electricity and has an orbiting motion at 10,000 Orbits per minute for a fast, smooth finish. Various grades of pre-cut sanding sheets are available to suit all types of preparation on vehicle bodywork. I've had this one for about eight years.

models are available, some with dust extraction facilities which in this day and age must be considered more environmentally friendly. You don't want all the dust from your plastic filler drifting away in the breeze, only to fall into your neighbour's fish pond.

All in all a very useful tool.

Putty Knife

The putty knife can be simple or complicated. I use an old butter knife. The knife needs to be thin, and flexible. Other than that, either buy a special one from your supplier, or raid the cutlery drawer!

Paint Scraper

Paint scrapers are available from household decorating shops and Do-It-Yourself stores. They consist of a wooden or plastic handle and a flat bade about 3 inches wide. They are essential for use with Chemical paint removers. The paint scraper should be kept clean and free from underseal. Most scrapers seem to end up with underseal on them, so if you use your scraper for removing old underseal clean it properly afterwards with white spirit or thinners.

Aerosol Paint

This might seem a bit unlikely in a section on tools. But as you will see in the following Chapters there are occasions when an aerosol of black paint can assist your preparation considerably. It is used to spray on a mist or guide coat.

The guide coat only exists to be rubbed off! If you cover the panel with a light spray of black, then rub all of it off you can be sure that you have sanded the entire panel. This very simple technique works wonders.

Masking Paper

Brown paper is recommended for masking as it is stronger than newspaper, and so can stand up to the considerable air pressure coming from the spray-gun. Also, thinners can inter-react with the printing ink on newspapers and give a messy finish.

Masking Tape

You will need two to three complete rolls of masking tape 1 inch wide for a respray on the average saloon car. Better safe than sorry. Buy

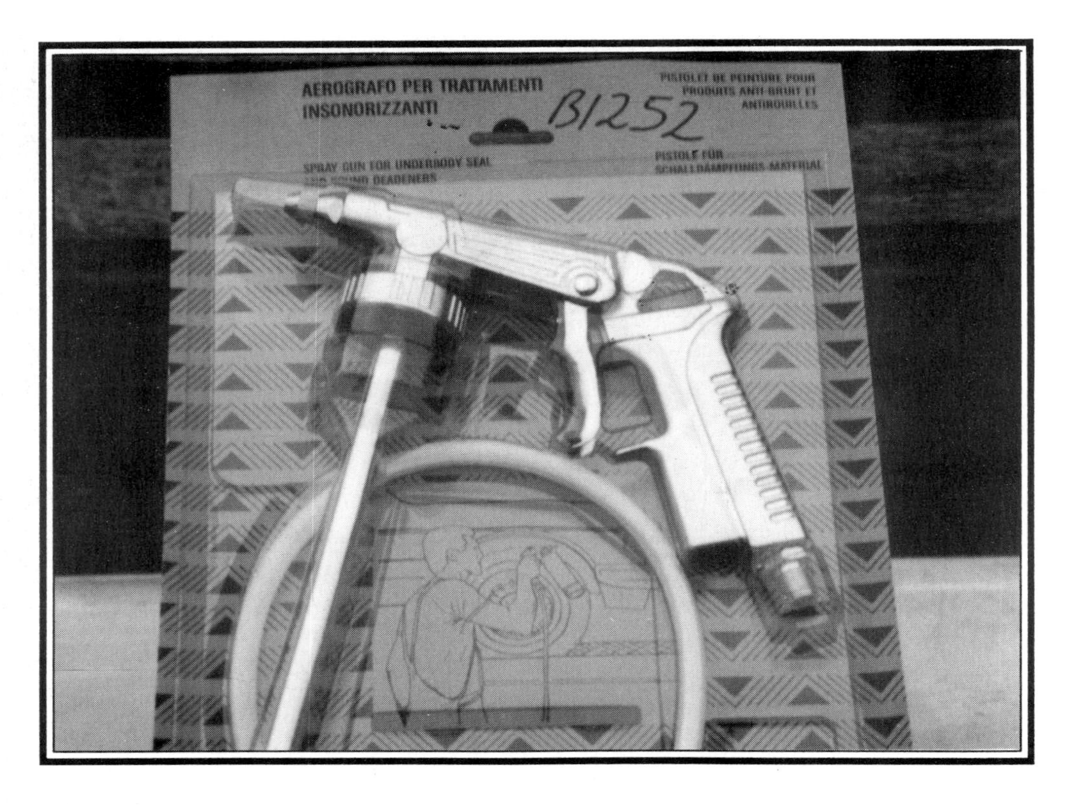

This special spray-gun is used for spraying underseal and sound deadeners. It does not require much compressed air and is quite cheap to buy. The flexible plastic hose can be used to spray inside sills etc. where there is normally no easy access. This type of gun is often referred to as a "Schults" gun.

three and have some left over. Buy two and run short on a Saturday afternoon -- the choice is yours.

Craft Knife

You will find a "Stanley" or craft knife very useful for cutting masking paper and tape. It is one of those tools which becomes essential as soon as you buy one. How did you ever manage before?

Plastic Bags

A number of large plastic bin liners can be used for masking wheels and tyres.

Plastic bags are also good for covering large areas quickly, such as covering up a dashboard when spraying round the windscreen aperture.

Plastic Bowls

A couple of plastic buckets or basins for the rubbing down water. Don't use your domestic washing-up bowl, as it will inevitably get scratched from being kicked round the workshop as well as having abrasive papers rubbed around inside it. They are cheap enough to buy, so get your own!

Washing Up Liquid

When using abrasive papers "wet" you need some soap or washing up liquid to help release paint particles from the abrasive paper. If you don't, the paper will clog. Fairy Liquid is good for this, but as it says in the adverts, you only need a little to do a lot. Remember to ensure that all detergents, such as Fairy Liquid are removed from the panels prior to spraying.

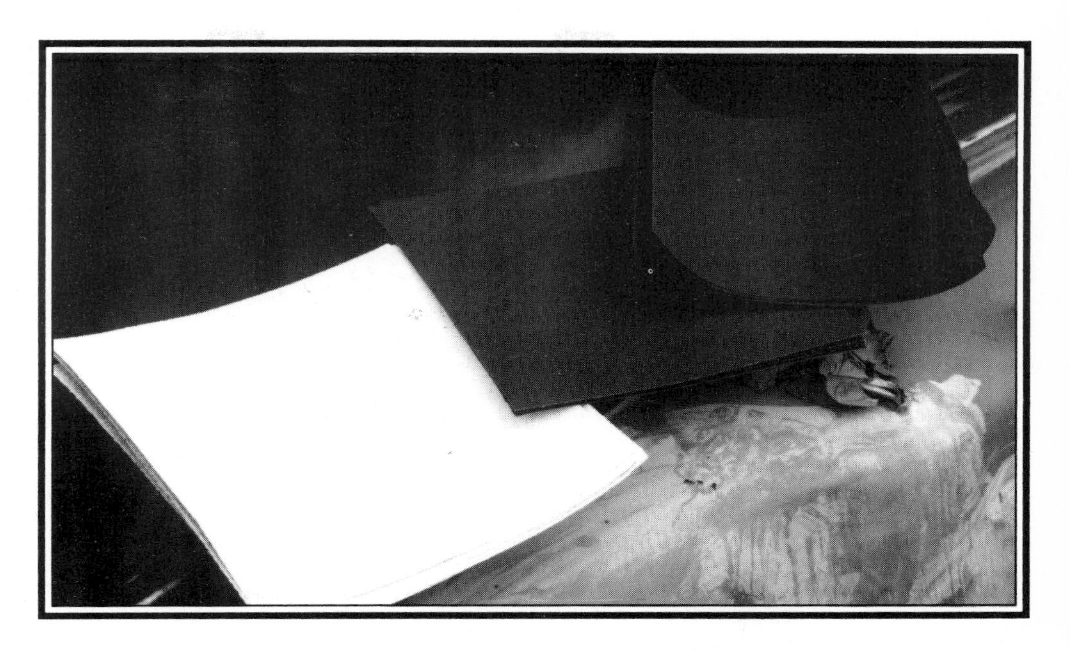

A selection of wet and dry abrasive papers. As you work a given surface use progressively smoother papers to achieve the best finish. 80 grit is very rough, 1500 grit is very smooth.

White Paper

A dozen sheets of white paper, A3 in size and some drawing pins. These are used to set up the spray-gun prior to letting it loose on the car. Adjust the spray pattern on a piece of paper pinned to the wall. See the information on spray patterns in Chapter Five.

Tack Rags

Tack Rags are special cloths impregnated with varnish. Their job is to pick up all traces of dust and other contaminants from the prepared panel, just prior to spraying. Various types are available depending on the paint system being used.

Preparation Time

The next Chapter begins to look at preparation of the surfaces to be painted. You cannot spend too much time preparing bodywork. Less than 20% of time is actually spent spraying when doing a respray.

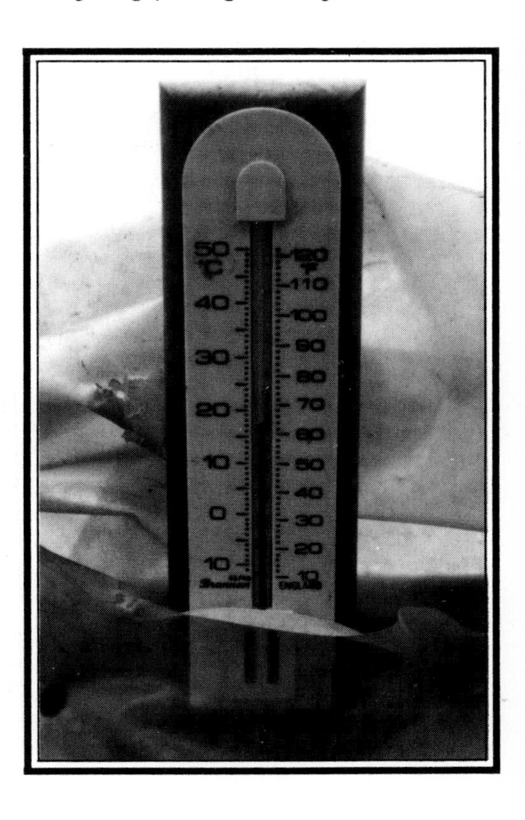

A cheap greenhouse thermometer is essential to keep an eye on the spraying temperature. This is a traditional fluid thermometer but hi-tech ones are available quite cheaply.

The bodywork has taken a long time to get to this stage! Two new wings have been welded on and the lower part of the front valance is new too. Although the wings look perfect it took two of us about four hours to make them into "perfect panels." Details of the work required are given in the following Chapters.

The Cortina (which also appears on the front cover of this book) is inside a "poly tunnel" which cost about 130 pounds. This tunnel consists of heavy polythene supported by tubular steel loops. As long as you remember it will be very hot inside the tunnel on a sunny day, it is a very effective, cheap and portable workshop cum spray-booth.

Preparation (1)

In the Introduction we indicated there would be two Chapters on preparation. This, the first takes you from the point when you decide to respray the car, to the point when you first think it is ready to be sprayed. The second preparation Chapter then focuses your attention on all the things you have missed and gets you into the fine detail of preparation which makes the difference between success and failure.

Remember, vehicle painting is 80% preparation and 20% spraying!

Identifying Paint

Before you start preparing your bodyshell, you must identify the type of paint already coating the panels. Take a rag and dip a corner into some thinners. Wipe the rag onto a small area of paint. You probably want an area that is not too important because you are going to soften or destroy the paint on the test area. If the paint is Cellulose, it will soften the paint and some will wipe off on the rag.

If the paint is synthetic it will "pickle". This means it will wrinkle up and look like the surface of a prune.

To test for Metallic paint put some T-Cut onto a rag. If you have Metallic paint some will wipe off and you will have a coloured area on the rag.

If the paint is Basecoat and Clear, all you will have on the rag is dirt. There will be no colour.

If the paint is Two-Pack the rag should make no impression on the paint.

Having identified the existing paint now is the time to consider what type of paint to spray.

Paint Codes

Manufacturer's paint codes were dealt with in Chapter One. Refer also to the Paint Codes section in Chapter Six, Mixing Matching and Miscellaneous and the following paragraph before you rush out and buy paint.

Before You Buy Paint

Check the colour code as described in Chapter Six but make sure you check the paint with a coded paint chip. Your paint supplier has these. If the car has had a respray it may not have been sprayed in the original colour.

Check all these points carefully before you order any paint.

Once paint has been decided upon the right thinners must be used and the correct hardener if using Two-Pack.

Rust Renovation

Let's look first at a bare metal panel with rust. Rust is similar to cancer in the human body. Once it is in there you can treat it if it is very, very light surface rust. If it has eaten into a panel then, like in a human body it must be cut out and removed.

Surface rust can be removed with an orbital sander and using an acid cleaner on a panel. Trevor says the best acid cleaner he has come across for corrosion is Dioxidine. However, there are many similar products on the market. Be sure to follow the manufacturer's instructions when using these products.

Prepare the panel by sanding it, then brush Dioxidine onto the panel. Mix the Dioxidine in the ratio of approximately six parts water to one part Dioxidine (or as directed by the manufacturer). Allow the panel to dry and the chemical to soak in for about five minutes. Wear a pair of rubber gloves and use wire wool to scuff the panel.

When scuffed all over, wash off with water, then dry with a clean rag. As quickly as possible after the panel has dried apply primer over the bare metal. Do not wait too long as air oxidises the metal back to its rusty

This Lancia Fulvia Sports has an alloy body. Terry Burville spent the best part of three days with an angle grinder and a 36 grit disk removing all the old paint from the bodyshell. This is known as "linnishing". Terry believes that chemical strippers get into every crease and fold in the body and are very difficult to remove. Chemicals NOT removed will damage subsequent paintwork. Although the linnishing process leaves the surface roughened, which will provide a good key for paint, an etch primer must still be used.

state almost before your eyes. Do not, for example, prepare the panel then leave it overnight without spraying on primer.

If the metal is slightly pitted with rust, the surface needs to be prepared a bit more thoroughly than by sanding. Sanding tends to leave the pitted areas with rust still in the pits. The best way to tackle a pitted surface is shot-blasting equipment followed by Dioxidine, which will eat into the rusted areas.

You must be very careful using shot blasting equipment. If the metal is not very thick to start with, plus corrosion has eaten into the metal it can become very thin. Holes could be "blown" in the metal by the action of the shot blasting equipment.

If the metal is pitted very badly and there are holes already in the panel then these need to be cut out and a metal plate welded in. If there too many areas to be plated then the old panel must be replaced with a new panel -- if one is available! This seems to be a good opportunity to recommend "Panel Craft", the sister publication to Paint Craft, which deals with car body restoration.

Before applying primer on panels we need to ensure that the surface has been prepared properly.

If there is an excessive amount of paint on the panel and it has started to crack, you need to strip back to bare metal.

This is very easy, as you brush a paint remover on, leave to eat into the paint for 5 or 10 minutes, then scrape off the resulting mess and wash the panel down with water.

You may need to repeat this process as often as required, depending on the thickness of paint on the panel.

For fibre-glass motors you need to use special fibre-glass paint stripper as solvents in

Terry has applied a layer of body filler on both corners of the lower, front valance. After allowing the filler to harden, he smooths the areas with a body file. This is like a cheese-grater and quickly removes filler in large pieces. After achieving a smooth finish with the file, various grades of wet and dry paper are used to further prepare the surfaces.

ordinary paint stripper are too strong and will soften up the fibre-glass material as well as the paint.

Once you have used paint remover on fibre-glass you need to wash it off, then leave the whole job to settle for as long as possible, certainly 4 or 5 days. The reason for this delay is that some solvents will eat into fibre-glass and they need to dry out before it is safe to continue with the painting process.

If you have repaired an area on the panel, sand down the area to be filled. Make sure you have bare metal surrounding the filled area. When you have finished the filler work properly, rub down with production paper -- use 40 grit if there is a lot of filler, finishing off later with 80 grit.

Sand down the edges of surrounding paint. This process is known as featheredging and it ensures you have a nice flat surface.

You'll need some Cellulose Putty* on top of the filler to cover scratch marks or pin holes which occur in the filler.

The Cellulose Putty* needs to be scraped on thinly, rather then put on in a thick film. Once done and prepared properly, sand down with a powered sander ready for priming.

If a powered sander is not available, be prepared to spend some time and effort with a sanding block.

If flatting wet and dry use 280 or 320 wet and dry paper.

If using 280 or 320 grit paper you must ensure that the area flatted with this coarse a grade of paper is going to be primed over.

PRIMERS

If you are spot priming certain areas you need to flat remaining paint with wet and dry paper.

Once the panel is ready for priming, you must choose the most appropriate primer. If

*Acrylic Stopper can be used instead - sinks less than cellulose.

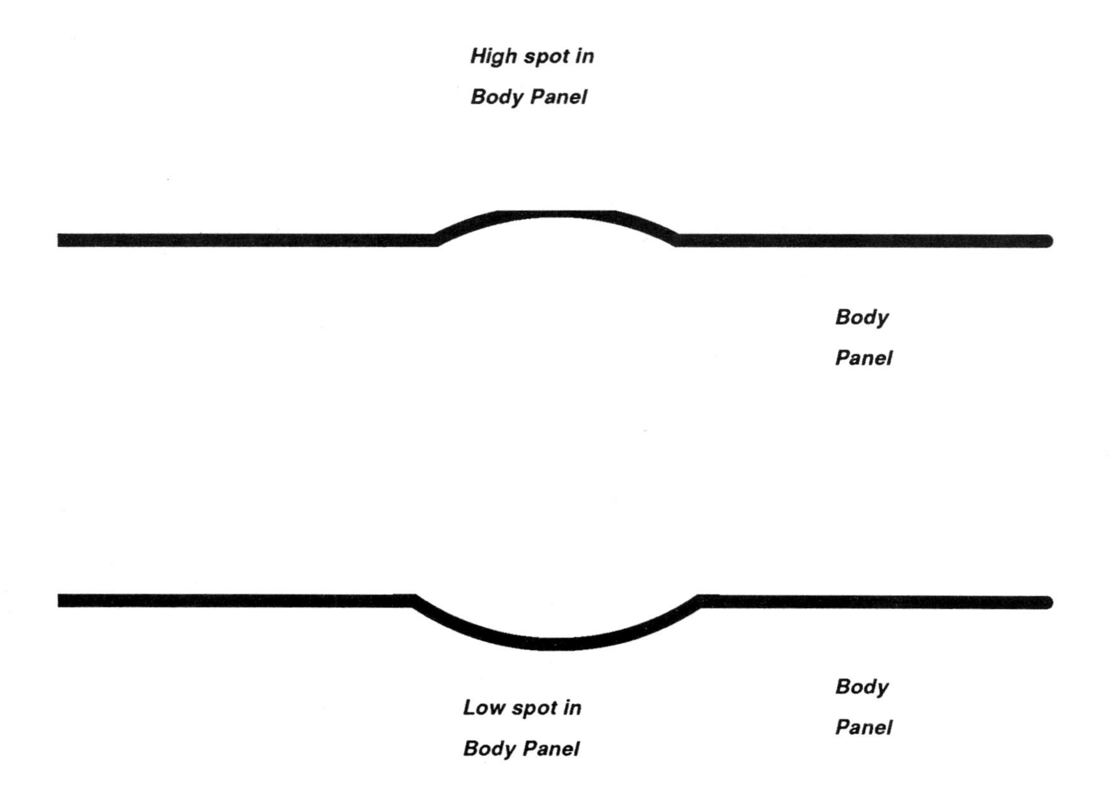

The low spot (lower illustration) can be filled with plastic or lead filler and smoothed flat ready for priming.

The high spot (top illustration) must be tapped down so that it becomes a low spot, which is then filled. The alternative is to apply filler and raise the level of the panel to the height of the high spot. Both methods can be used, but it is cheaper and quicker to deal with the high spot ONLY rather than work on the entire panel.

you are spraying over fibre-glass, aluminium or bare metal you must start off with a self-etch primer. This is certainly essential for aluminium but not quite so necessary for other materials, as most primers available nowadays have self-etching properties. You must use a pure self-etch primer on aluminium otherwise flaking will occur.

Self-Etch Primers

There are basically two types of self-etch primer;

"Long life" where you mist coat two coats of self-etch primer plus another appropriate primer for the top coat, or Two-Pack self-etch non-Isocyanate filler. This is a filler and etch primer combined. so you can miss out primer stages after spraying the etching primer as it is all done in one. Just flat the primer and carry on with the top coating.

Other Primers

There is quite a choice of primers available following the self etching primer and the decision as to which to use is governed by the type of paint being used.

Two-Pack primer is the best all-round primer, as it gives such a high build and goes off so hard, covers most scratches and does not sink. However, if you do not have an air-fed mask you are back to Cellulose primers.

You need to build up four or five coats, leave for about a day to dry out thoroughly,

Using a two-by-two sanding block cut from some scrap timber, Terry Burville uses some 60 grit paper to smooth the plastic filler. Ripples and dents in the wing caused by welding and careless handling are corrected by this process. It can take two or three hours to prepare a brand new panel in this way.

then spray on a guide coat followed by flatting off.

Then spray two or three coats again, but a lot thinner so that you have a nice smooth finish and don't have to do so much flatting down. The first four or five coats give the build, the second two or three gives a smooth finish, ready for top coats.

Repairs

Most of you will be spraying a "classic" car, that is one which may be twenty or more years old and has had some repair or renovation work carried out on it. This repair work must be finished off correctly as a respray will not hide shoddy repairs. In fact a respray will emphasise shoddy repairs!

Any welding work on the body will have to be ground down smooth, either with an angle grinder or some other sort of power tool. Having got the welds down to a smooth finish, next consider if any of the welds are forming a high spot on the panel. A high spot is any section of the panel which is raised above its surrounding area. If a high spot is found, you need to do some hammer and dolly work to flatten the high spot. Remember that low spots or depressions can be filled with body filler or lead, but high spots are a problem. The best you may be able to do is hammer a high spot down so that it becomes a low spot, then fill it with plastic or lead body filler.

Spray Putties

You can use Acrylic Putties which can be used over the top of Acrylic and over Cellulose.

You must be very, very careful if using Plastic Padding Acrylic Stopper over the top of One-Pack Acrylic as they do tend to sink and cause you problems.

Chapter Three

If using stoppers you must make sure that the layers you put on are very fine and thin. Unlike filler, where you are putting a fair bit on to fill, stoppers are only used for very small pin holes or light scratches so you don't need to put a big build up on.

If you need a little bit more then, yes, go over it a second time but don't try to fill it all in one go with stopper.

Stopper is only there for small scratches and dents -- not to fill!

Dealing with Dents

Dents can either be beaten out with a hammer and dolly or filled with body filler. Remember that there is an optimum point for working out dents. You can spend an hour on a dent and get 90% of it to the correct contour. That remaining ten per cent may take another two or three hours to see little improvement. Don't spend unnecessary time on dents. Do your best to bump them out then fill them. Spend your time more effectively by finishing off the filling operation and getting the filled surface absolutely smooth.

Removing Underseal

Underseal can be removed with thinners or petrol if the underseal is still soft or wet, but not if it is hard. Hardened underseal needs to be heated or scraped -- both methods being a bit messy but unavoidable.

Removing Paint

There are several way to remove unwanted paint, perhaps the most ferocious is to use a chemical remover such as "Nitromors". But before we look at these techniques let's try to find out why we need to remove paint.

The "back to bare metal" respray often described in the monthly magazines may be the ultimate way to carry out a respray, but for the amateur there may be no real necessity to remove all the old paint. It depends on several factors.

For example, if you are going to change the type of paint used, then the original paint may not be compatible with the new paint -- so the old paint has to be removed.

If the paint is in very poor condition again you may wish to remove all of it. The classic example of this is a crazed or broken up paint surface.

Chemical Removers

NOTE: Use special chemical removers for glass-fibre bodywork.

Most chemical strippers come in a tin and are in heavy liquid form. They are brushed onto the paint to be removed and left for a certain time. Refer to the manufacturer's instructions. Do not rush this process as it is no use trying to get the old paint off before the chemicals have done their work.

All rubbers, such as door rubbers or windscreen surrounds need to be removed or masked up, otherwise the paint stripper will perish the rubber. When stripping a panel or vehicle it is best to lay down some one inch masking tape on the edge of the panel. This stops the paint stripper creeping round the edges. If you don't tape up the edges and paint remover does get behind the edges then the door shuts or inside of the boot or bonnet will need respraying. This will cost you a lot of extra time and money.

Do a small area at a time, (such as a panel) then when that is stripped, go on to another area. For example you may wish to strip the bonnet in one session, and the following evening the boot lid and so on.

Generally removers need half an hour to do their work, after which time use a flat paint scraper to remove the paint from the metalwork. It should come off in strips or layers. Remember if there is a lot of paint to be removed you can put on another coating of remover after the top layers have been disposed of.

All this takes time and patience.

Once all the paint has been removed, you MUST kill any remaining traces of the paint remover from the bodywork. Refer to the instructions on the paint remover tin. If you don't remove all traces when you come to spray on the new paint you will get blisters and other problems.

Just before you actually spray the first new paint on, you would be wise to blow over the panels with compressed air. Get the air into

The opposite front wing to that shown on page 45. You can see the outline of the filled area towards the front of the wing and round the headlamp mounting. As most people look at the front of the car when judging a painted finish, it is vital that your preparation of front wings, bonnet and front valance is perfect. This wing needed about two hours of work to get it perfect. Look at the finished job on the front cover of the book.

This body filler is specially formulated for professional use. It is very good for "featheredging". Before applying this material you must remove loose rust and any damaged paint, and ensure that the surface is free from grease and dirt. Follow the instructions on the tin regarding mixing of hardener. This is a professional workshop's 7Kg tin.

all the little nooks and crannies which may hold traces of chemical stripper or water. You will be amazed how much you dislodge with a high pressure air line.

Dry sanding

Dry sanding is the process where you use either a mechanical sander, or just a sanding block and abrasive paper.

There is no particular problem sanding this way, but it is generally reckoned that you get a better finish if you sand "wet". This means of course that you cannot use any electrically powered sanding tools when sanding "wet".

If using a power tool for dry sanding be very careful not to gouge deep scores in the existing paint. It is all too easy to pick up a power tool and remove some paint, but one slip of the wrist and you could dig out a deep score which will take hours to repair.

Always use a sanding block. If you just rub the abrasive papers with your fingers you will get a couple of nice grooves where your fingers dig in and apply excessive pressure on the panel.

Sand blasting

Sand blasting can take several forms and can be very effective for removing large areas of old paint quickly. However, there are snags.

If you use a grit blasting technique where small grains of grit are blasted against the painted surface, you will have a never-ending job of getting rid of that grit! It will get everywhere and will resist all attempts to brush it away. Just when you think it is all gone, apply some compressed air to the gritted area and it will all come out to annoy you. Think what that will do to the finished paint job...

About Abrasive Papers

Abrasive paper, or wet and dry paper, as it is most often known is graded according to the size of the grit on the paper. The higher the number the finer the grit used.

15 to 40 grit used to remove paint and rust and are often used for grinding welds.

40 to 80 grit used for removing plastic body filler and body solder.

80 to 120 grit used to smooth the finish on body fillers. Good for feather-edging. Also used for removing heavy layers of paint, surface rust and for general preparation of metal surfaces.

120 to 220 grit used for roughing old paint prior to re-coating. Also suitable for smoothing out scratches caused by rougher paper.

220 to 400 grit used to prepare the surface prior to painting. Wet sanding often done with 360 or 400 grit.

400 to 600 grit used for final preparation and intermediate finishing during final paint coats. Good for Synthetics.

From **400 to 1500 grit** are usually described as ultra-fine papers for smoothing to a very high standard of finish. Abrasive papers at the higher end of the grit scale actually polish the paint being so fine. 600 to 800 are used for Cellulose or One-Pack Acrylics. 800 to 1000 is good for Two-Pack.

OPEN COAT or OPEN GRAIN

Used for paint removal and have less tendency to clog with removed material.

CLOSED COAT or CLOSE GRAIN

Used for final smoothing, often with water and/or a lubricant such as Fairy Liquid where less material is to be removed.

Wet Rubbing

Rubbing down with wet and dry paper is the most common method of removing unwanted paint. It is the best technique to use on a primed surface prior to putting colour on. You can also flat dry, but this sometimes clogs the paper and causes scratches.

Wet rubbing can be extremely time consuming and make your arms ache. All the effort will be worthwhile at the end of the job.

One tip is to use warm water. Much more pleasant to use, and you can also add a drop of Fairy Liquid to help the passage of the paper over the surface and prevent the paper clogging with removed paint particles.

Glass Fibre Preparation

Depending on how far you have to go, you should really use a fibre glass stripper. Any other stripper, such as a stripper for metal, can be too coarse and cause damage to the fibre glass by softening it. With fibre glass stripper, paint it on and scrape it off. Don't use too sharp a scraper or you risk digging into the fibre glass.

Once you have scraped off the stripper and the paint, and washed it off you MUST leave the car standing for four or five days (a week is better!) to allow any acids to evaporate from the fibre glass surface which may be slightly soft. If you don't do this you may find that when you prime the surface you will get a reaction between the paint, the soft fibre

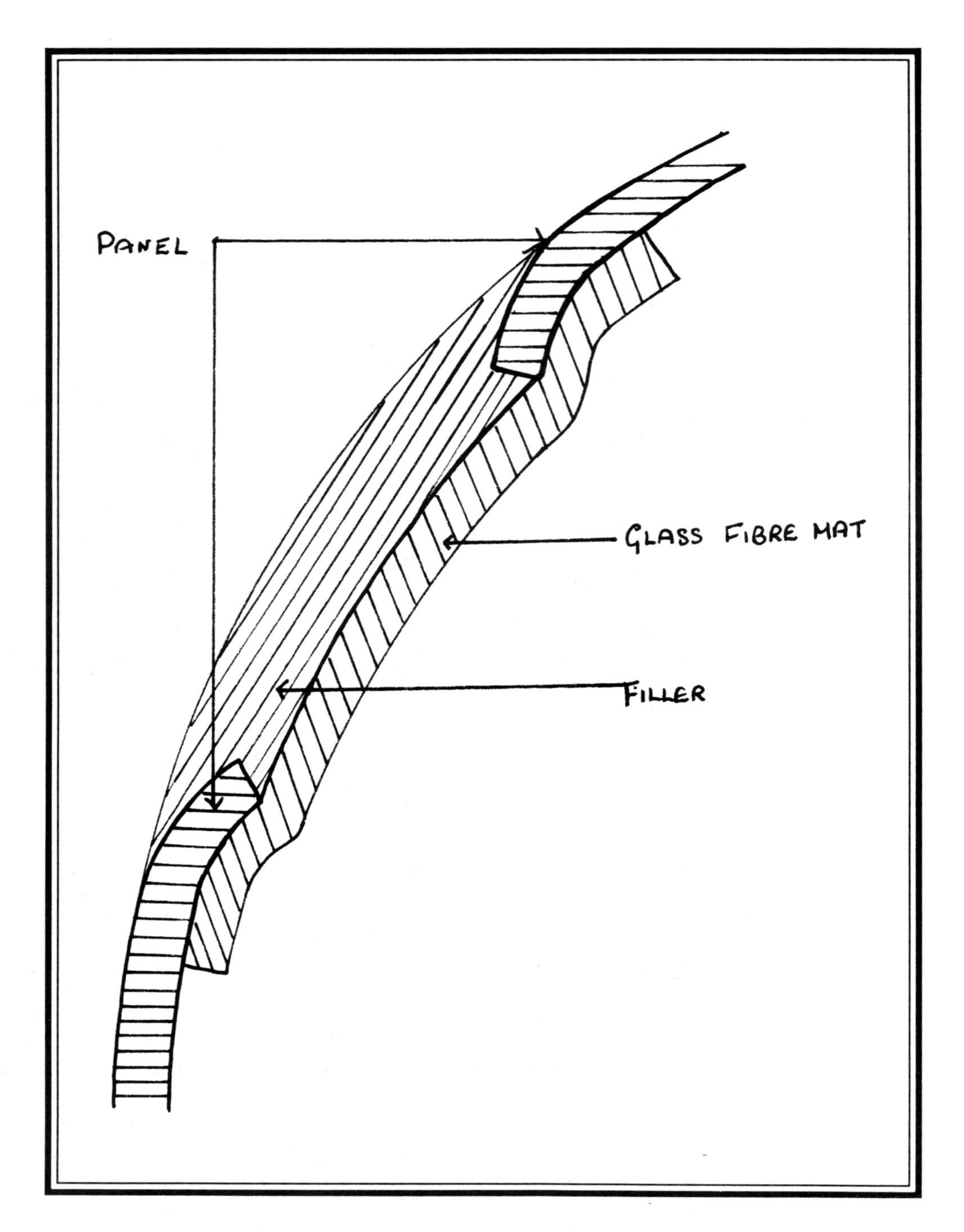

If there are holes in the panel you are working on, you may be able to repair them with glass-fibre mat and filler. You can see the technique from the above illustration. You may need to support the glass-fibre mat until the resin sets. There are full instructions with glass-fibre repair kits.

find that when you prime the surface you will get a reaction between the paint, the soft fibre glass and any residue from the stripper.

Once you have prepared the fibre glass and rubbed it down, you are ready for priming. Depending on what primer you are going to

The wings and top of the front valance have been coated with a high build primer, sanded down and coated with grey primer. Terry Burvile has applied Cellulose putty to the lower valance (you can see the marks) and is using the angle grinder to smooth off the lumps.

follow up with, to start with you want an etch primer on bare fibre glass.

This can be a long life etch primer, or self-etch primer comprising etching primer plus primer.

The long life primer process is two coats left to dry for about an hour. Do NOT flat down. Follow this by spraying Cellulose or Two-Pack primers.

If you are using a etch primer with primer added (like a Two-Pack primer) this will do the job of etching it and also building the primer up. Leave the primer for a day or two (longer if possible) then flat off knowing that there should not be any problems.

If any problems are going to occur they will appear in the primer within a couple of days.

The next step is top coating, depending on whether you are going to use Two-Pack, Cellulose or Synthetics. Cellulose is not used much on fibre glass as the finish goes very hard and brittle. This is no good on a fibre glass vehicle which tends to vibrate and crack the paint. Two-Pack is recommended for fibre glass.

Spot Repairs

If you are doing a repair on a panel, grind down to bare metal, tap out any high spots prior to adding filler. When the panel is as smooth as possible, add the filler. Next, block down the filler to flatten and smooth it.

Don't try to cover paint with the filler. Grind away a fair area round the repair so that there will be bare metal round it. Then put filler on, but don't go to the edge of the paint. Almost to the edge, or if you want you can go just to the edge as you are going to sand it away anyway. Then sand down the filler until you get your repair properly, then get an orbital sander or a rubber block, and rub away the edges of the paint as you will have deep scratch marks where you went over the area with a grinder. Get rid of those and feather it out nicely so that there are no

TOP: Preparation is going to be difficult on this 1963 Ford Capri. All the lamp fittings will have to be removed before the flaking paint can be sanded smooth. At the bottom of the headlamp panel there is a seam which is notoriously difficult to seal and keep free of rust.

BOTTOM: You can see surface rust plus a lot of dents on the rear quarter panel of the Capri. The lower six inches of the panel was replaced a few years ago but the job was never finished properly. It is a lot better than it looks, though! Three or four hours should put it right.

If you look carefully you should be able to see lots of imperfections in the paint on this Capri bonnet. Looks like it was put on with a trowel! There is a lot of paint on this panel and the best course of action is to strip it back to bare metal. We can confirm that when this was done there was very little repair work to do - just a small circular dent in the middle of the bonnet.

scratch marks and no edges to feel. It must be nice and smooth.

Then spray on primer followed by Cellulose stopper if required. The other way to tackle this is do the filler work and add Plastic Padding Stopper over the top of the filler. Sand that down and then you are ready for priming.

Once you have got to that stage, you should not have any problems. You should be able to put the top coat on without any "Mapping".

The problem known as Mapping is if you fill over the top of paintwork, the filler sets hard and then when you come to spray primer or top coat on, the solvents from the thinners tends to creep under the edges of the filler. This gives you what is known in the Trade as Mapping or Ringing. The outline of the repair tends to show through the top coat. You can sometimes get rid of this by rubbing down with 1200 or 1500 grade paper and polishing. If it is too bad, you really need to block back down, prime over the section and re coat in colour.

If you have primed the panel and are not going to use Plastic Padding over the top of filler, (in other words you are going to prime then use Cellulose stopper) make sure you re prime over the Cellulose stopper. Trevor has known people to fill small pin holes with Cellulose Stopper, block it down leaving just a minute bit of stopper the size of a pin head. They think it is so small that they do not bother to prime it. They put colour straight over the top, but they get mapping edge coming back through so you still see the small pin head repair.

It is ***essential*** that the preparation is right before you put the top coat on because all you are going to do is waste money putting top coat on when you are going to have to re coat in primer and top colour again. **This is a MUST!**

Chapter Three

Spot Repair on a Bonnet

We would say that the bonnet is probably the hardest place to get rid of a fade out technique or a blow in as it is called, mainly because it hits the eye.

The roof is normally eye level but the eye does not seem to catch on the roof so much as the bonnet. The bonnet is the most difficult part as it is a fair sized area and it is always better to spray the bonnet and if you are worried about a colour match spray the bonnet and blow in the scuttle panel and over the two wings rather than try to do the fade out on the bonnet.

It is possible to do a fade out on a bonnet if you are a professional but it is difficult for the amateur, mainly because the preparation has to be so good. If you are going to do a local repair what we need to do is T-Cut the panel first of all and bring the colour back as close to its original colour as possible.

Then prepare the area you are going to spray. If you have filled it and primed it you need to rub down the primer and T-Cut away any overspray and then Scotchbright the panel over then do your fade out. If you are using Cellulose then do a fade out, blow in the edge with thinner, then T-Cut up.

If you are using Two-Pack it is just as easy to fade it out and then lacquer over the top of it. You are then lacquering the complete panel even though you faded out a on just the repaired section.

Preparing New Panels

You know those "new" panels you fitted? The ones which came finished in a nice brown works primer? Well, I'm afraid if you want to do the spraying job properly you will have to give them a good flatting down to provide a key for the new paint.

Opinions differ on this, but since we are trying to encourage good habits in this book, (rather than how to take short-cuts), we have to flat the works primer. Strictly speaking it is not really a primer, more a sealer. Many garages and professional body shops will spray primer straight on top of this protective sealer, but as we said a few paragraphs ago the sealer's surface is too smooth and has to be roughened with wet and dry paper to give a good key. Use a 280 or 320 grade paper and some soap and warm water. To do a thorough job you will need to spend about an hour on an average wing panel. This works primer is a lot easier to remove with an orbital sander and a 180 sticky disc.

You don't want to spoil the finished job for an hours work, do you?

Plastic Body Filler

Plastic body filler is ideal for the restorer, being cheap to buy and easy to work. For these reasons it is also used by the professionals! The use of lead for filling dents has long since passed, with only dedicated craftsmen still using lead. If you shop around you should be able to find someone using lead if that is what you want on your car, but there is absolutely no reason why plastic shouldn't be used instead. If you wish to use lead there are detailed instructions on the following pages.

If you intend to use plastic body filler, then buy a large tin. Commercial size tins are available, but you may have to go to a professional's supplier.

Having got your tin of plastic, what do you do with it? Well, the first thing is to prepare the metal to be filled. This may involve hammer and dolly work, or just bashing out a dent with a small hammer. Either way, remember not to spend excessive amounts of time on a dent. Most of a dent can be improved within an hour or so. Any more time spent may not be justified. You are going to fill anyway, so why not a little extra filler.

It is essential that metal is free from paint and that means grinding or sanding off any paint in the damaged area. Make a thorough job of this as plastic prefers to bond to clean steel, not painted surfaces.

Having repaired the dent and ensured that the area to fill is free from paint, oil and other contaminants, mix your filler on the plastic lid which is provided. (You will find a plastic lid, then a tube of hardener, plus a flat piece of plastic to use as a spreader. There are usually some instructions in the tin as well).

Mix the hardener with the resin as recommended by the instructions. We cannot give hard and fast rules about this as it does vary slightly depending on the air temperature you are working in. The chemical reaction of the two ingredients being mixed gives off heat. If you are working on a very hot day, you can use less hardener. If working on a very cold day, then more hardeners will be required. Strictly speaking the correct amount of hardener should be used whether it is hot or cold. Too much hardener can sometimes cause staining in the top coat and can also slow the paint drying process.

Mix the two ingredients thoroughly. The mix should change colour, so make sure the colour is even throughout the mix. Only mix up as much as you can use in 10 minutes, for after this time the mixture will have hardened so much as to be unusable.

Spread the mix into the area to be filled using the plastic applicator. Make sure there are no air bubbles in the mixture, and try to get it smooth and level. If you don't then that job will be done by a sander, or sanding block and muscle power. Obviously the smoother it is to start with the less work will be needed in rubbing it down when dry.

Wait until the mix is dry, this will be within 10 to 20 minutes depending on how warm it was and if you got the mix right. As soon as the mixture is hard, begin to rub it down.

If you find that one area is still too low, in other words needs more filler, don't worry. Just mix up some more and add it to the problem area. There is no problem about building up layers of filler, as long as you do not overdo the process. Three or four layers is probably a safe maximum. Always aim to put a little too much on, then you have to spend the time sanding it down smoothly.

When it is smooth and you are satisfied with the repair, it is then ready to be prepared for spraying.

This mainly involves featheredging the repair so that it blends smoothly into the surrounding paintwork. That word, "featheredge" will keep cropping up. Note the definition, "blends smoothly". That is the key to the whole repair and hence the quality of the respray. Take time and trouble to get these plastic body filler repairs right, and they will vanish completely when the paint is applied. Skimp on the job and they will shine like beacons through even the best spray job. The quality of the respray is in the preparation.

Body Soldering

Body soldering, lead loading, and leading are all terms used to describe the process where lead is used to fill dents instead of plastic body fillers.

The lead technique is much older than the plastic filling process which is a relatively recent invention. Solder techniques go back to the 1930s but have been largely replaced by plastic during the last 20 years or so.

Lead is still favoured by "old time" craftsmen, and many younger restoration enthusiasts are returning to the lead techniques. Several specialist car restoring catalogues now offer body lead kits which contain all the necessary items to allow the amateur to learn the lost art of body soldering.

Lead is also the ONLY filler which can cover holes and is impervious to water. It is possible to fill quite large holes by placing a material which cannot be soldered, such as aluminium or wood, behind the hole. Lead the area as described later in this section, then withdraw the backing material to reveal a perfect repair as carried out by real craftsmen.

Car body solder may be slightly different to plumber's solder, in that the ratio of tin to lead is different. In car body solder the ratios are usually about 70% lead and 30% tin. You can use plumber's solder without any worries, and you may find it cheaper to buy.

The solder must wipe smoothly and not crumble or break up, and must have properties which allow it to be reheated several times during the wiping process.

The user needs body soldering blocks, usually made of wood and in various shapes depending on the jobs being tackled. These blocks are often referred to as bats or paddles. A special oil such as tallow is used to prevent the solder sticking to the paddle. Apart from the solder itself you need a source of heat, such as a welding torch or propane torch, some tinning material (often a paste) and a brush, and some clean rags. When the soldering is complete you will need body files or sanding materials to smooth the solder to the required contours.

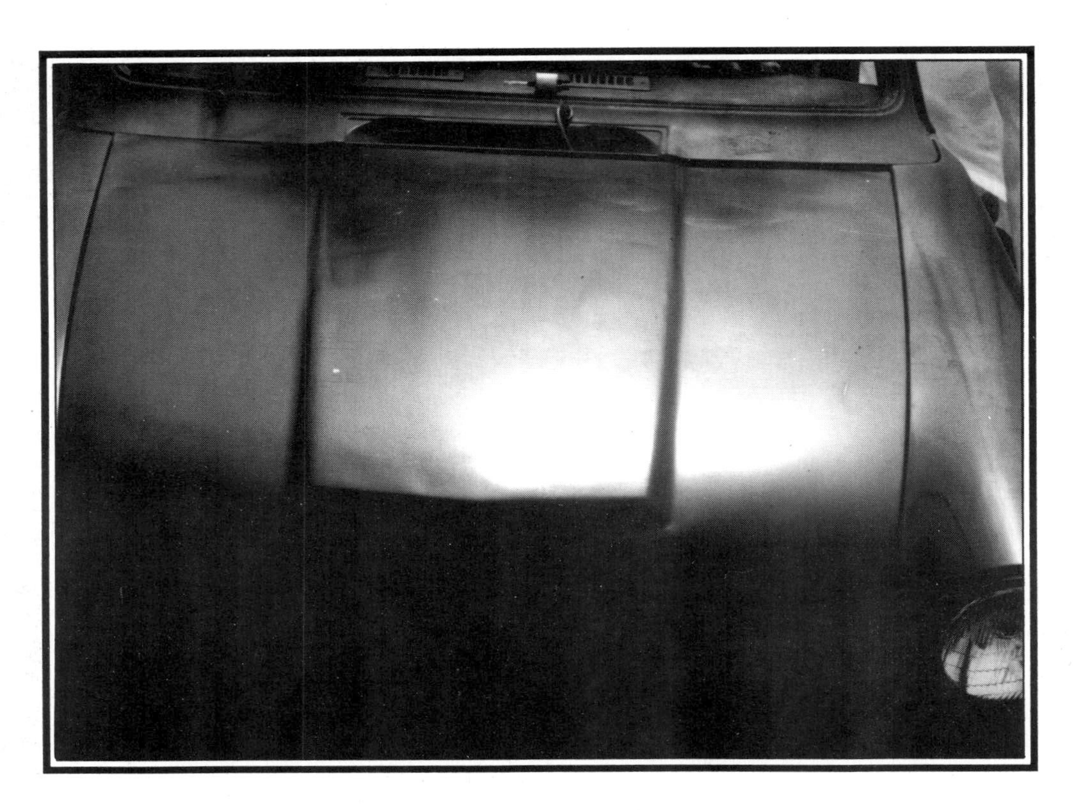

If you look closely at the rear of this bonnet you will see several areas where filler has not been flatted properly. The problem is, this is Tommy's second attempt at the bonnet! Black shows every little (or large) imperfection, so the only solution is to go back to bare metal and start again. You cannot spend too much time on preparation.

How to use Body Solder

Follow these general steps for perfect body soldering. Remember if you buy a body soldering kit make sure you get instructions with it. If you are soldering for the first time, try to tackle a horizontal surface rather than a vertical one. You'll find it much easier!

1) Prepare the area to be filled. Remember to check for any high spots, as you cannot fill high spots. They must be knocked down to be low spots, which can then be filled. Although this sounds obvious it is surprising how often you forget to check this, and only find out when you have applied the filling material.

Clean the area to be filled thoroughly. Remove old paint, rust, and dirt. An angle grinder is ideal for this task. The cleaner the surface the better the solder will adhere to the metal.

2) Apply tinning paste with the brush so that the required area plus a little bit more is covered.

3) Use the heat source to warm the tinning paste until it starts to become fluid. If you are using a welding torch adjust the flame until it is a "carbonising" flame with a very soft heat. It should turn from grey to a brown colour.

If you overheat the tinning paste it must be removed, as solder will not adhere to burnt tinning material. One clue to the tinning material being too hot is that it often turns black or blue.

4) Wipe the tinning solution with a clean dry rag. Professionals use a cloth called a moleskin. You should be able to get one from a plumber's supplier. And in case you are worried it is not really the skin from a mole!

Ensure that the area to be soldered is covered by the tinning solution.

5) The next step is where the skill lies. You may need several attempts to master it.

Here is the problem which has shown through the final colour coats -- see the photo on the previous page. Towards the bottom is a line where spot welds have fastened the outer skin to the frame. This repair now needs careful filling and flatting down. Remember you can wipe a damp rag over problem areas to see what they will look like when the final colour coats go on.

Holding the heat source in one hand, hold the stick of solder in the other. Apply the heat so that part of the tinned area, AND the end of the solder stick is heated equally. As the end of the solder starts to melt, touch it to the tinned area which you are heating. A piece of solder should now be sticking to the tinned area. Some experts recommend a "twisting" action to detach the heated end of the solder stick. Repeat this process until you think you have enough solder on the panel to complete the filling job.

6) Dip your wooden paddle in tallow. Keeping the flame playing on the solder, use the paddle to spread the solder over the required area. This technique again needs

Our ace sheet-metal man, Terry Burville has made up new body panels for this 1932 Hornet Special. Just in front of the door is the new scuttle panel made in three pieces -- top and two sides. You can't see the join! The new panels have been sprayed in a protective primer which will be removed as part of the preparation process prior to painting.

some practice, or you will find bits of hot solder dropping off onto the floor, or it will not spread because it is not warm enough. It is a bit like spreading butter. If the butter is cold it drags on the bread. If the butter is just right it spreads. If the butter is too warm it drips off the bread and makes a mess! You can also use the moleskin for this spreading.

7) Keep the tinning material warm as well to ensure good adhesion between solder and panel.

8) When you are satisfied with the solder spreading, use a body file to sand the area to the required contour. Do NOT, repeat NOT use power tools when working with solder. The lead dust produced is very dangerous to your health.

9) If required use fine grades of abrasive paper to get the required finish. The next bit will surprise you, but very often an experienced body repair man will finish off a body solder job by applying some plastic filler as the finishing touch to the job. Of course, in the old days the craftsman would continue to work the solder until the final, perfect shape was achieved. Nowadays, time is money and it is often quicker to apply some spots of plastic to complete a lead job.

In the last few paragraphs we have briefly described the art of body soldering. It takes time to become proficient and this is one of the reasons for the introduction of plastic body fillers. By all means attempt to learn the art, but don't forget you could do the job quicker with plastic filler.

Remember that solder is waterproof. Plastic filler is not. So if you choose to use plastic body filler, remember to protect it with a coat of primer as quickly as possible.

Drilling Holes

If you have fitted new panels to your vehicle, the chances are that you need to drill

Just a reminder that every vehicle has an underside which needs to be painted and protected. Those chassis box sections are good at rusting out and should be protected on the inside either by spraying paint, underseal, Waxoyl or all three. The large flat areas on this Cortina are being brush painted with Hammerite.

some holes in it. They may be needed for wing mirrors, door mirrors, badges, and so on. You MUST drill these before you paint the car. If you try to drill these holes after painting you will at the very least chip off some paint round the hole. At worst you will be inviting rust in, and within a year the hole will be surrounded by rusty metal.

Why? Because the hole you drill after painting will have paint on the upper surface and on the lower surface, but will not have any paint on the inside of the hole. In other words, however thick the metal is, that amount of metal will not be painted. So you are actually writing an invitation for rust to attack your new panel.

There are ways to drill holes without damaging paintwork, perhaps the most common of these involves you sticking a piece of masking tape over the area to be drilled. Next mark the spot to be drilled with an X in marker pen or pencil. Gently apply the drill bit to the marked area and drill. The masking tape will help to protect the painted surface from the drill bit as it attempts to cut into the steel.

This method helps, but does not eliminate the problem outlined in the previous paragraphs. Always, always, drill holes before painting. That way the hole is fully protected by the paint.

Of course, it goes without saying that once the hole is protected with paint you should be very careful to preserve that paint when fitting the accessory! Careless fitting can ruin all your good work and forethought.

REMOVE OR MASK?

The next problem is to decide whether to remove items of trim or to mask then. This can be a difficult decision. Most of the books say that it takes as much time to remove, say a bumper, than it does to mask it properly. However I wonder how many of these

Work on this wing is nearly complete. Remove all the filler dust from the top of the inner wing. Before priming, wash the wing thoroughly with water, dry it off thoroughly with rags and/or compressed air, then wipe it over with a clean, fluff-free rag and de-greaser or panel wipe.

professionals have struggled with a twenty-five year old bumper bolt which may never have been removed since the day it was fitted at the factory. You will have to use your common sense here. Try to follow the advice that it is BETTER to remove than mask, but if you find something stubborn like that bumper bar nut, you will have to make your own decision. You may be able to hacksaw through the bolt, but beware doing damage to either the chrome bumper or the surrounding bodywork.

If you have carried out a major rebuild then most of these troublesome parts will be removed anyway. Door handles, bumper brackets, wing mirrors, aerial and so on.

Remember if spraying inside the engine bay that you will need to remove items like the heater box on some cars, brake and clutch master cylinders and so on. Other points to be considered are bolt holes which will need to be masked off on the inside of the car by using sticky tape. If you miss one of these empty bolt holes you will be surprised how much paint will find its way onto seats, carpets and so on.

So you think its ready to spray, do you?

Just when you think you are ready to spray, along comes an expert to look at your work and what does he find? Inadequate masking, poor rubbing down, repairs not adequately finished, welds not smoothed off and so on.

In the next chapter we continue the preparation process to make just another home respray into a professional finish you can be proud of.

Preparation (2)

In the opening Chapter we indicated there would be two Chapters on preparation. This, the second, focuses your attention on all the things you have missed and gets you into the fine detail of preparation which makes the difference between success and failure.

Remember, vehicle painting is 80% preparation and 20% spraying!

Is it ready to spray?

No! If you want to spray the car now, you can but the results will be disappointing. Professional body repairmen reckon that, as a rule of thumb, you need to spend five or six hours preparing ***each*** panel prior to spraying. This could easily mean in excess of 80 hours work on a saloon car. The spraying could be done in a few hours, but the preparation could take weeks.

By this measure you spend five hours preparing one door, or five hours on the bonnet, or on a wing. Of course, times can vary and there will be situations when one or two hours will give you a perfect panel, while other times you may have to spend seven, eight or more hours on a panel.

For example, if you have fitted a brand new factory-fresh front wing to your car, you may be excused if you think the wing is perfectly contoured and ready to spray. Not necessarily so.

Get a wet rag and wet the surface of the wing. If the light is good, you should be able to look along the wing and see any ripples highlighted by the wet film of water. What you are seeing now is what you will see when the final colour coat is sprayed on the wing. To put right a few little ripples may take all day.

First mark with a pencil the areas you think need to be filled. Next roughen ("linnish" is the correct professional term for this process) the area with a coarse grinding disc if using an angle grinder, or a piece of coarse (40 grit) production paper. This will prepare the area for filling.

Now mix up some body filler. Mix up more than you think you will need. One thing which separates the amateur from the professional is the approach to using body filler. The professional mixes up a large amount and smooths it on over the area to be corrected. The amateur mixes up a little golf ball size quantity and spreads this over the same area. The amateur then satisfies himself with half a job, or has to go and mix some more. Meanwhile the professional is using some coarse production paper, or a body file, to start to shape the filler.

This sanding process can take a long time, even with mechanical help. You may not even get it right first time, and have to mix another quantity of filler to be applied over the top of the first. However, after a time you will start to see the filler taking shape on the wing. Keep working the filler until you have a smooth, shapely wing.

Then spray on a layer of primer. Allow this to flash off. If there are any imperfections, use Cellulose putty and a flat blade to work the putty into the scratches or holes which are highlighted by the primer. Follow the instructions on the tin, allowing plenty of time for the putty to harden. Don't be in too much of a hurry to sand the putty as you run the risk of lifting it all off again!

When the Cellulose putty has hardened, and you may need to allow a couple of hours for this, sand the putty down with 240 grade paper used wet.

When the putty has been sanded smooth, clean off the panel, wipe down with spirit wipe or panel wipe and spray on another coat of primer. When this primer has flashed off, it is time to put on a guide coat, often known as a mist coat.

Guide Coat

All you need to apply a guide coat is an aerosol of black paint which you spray on until you have a speckled appearance on the panel. This assumes you have the normal grey (or yellow) primer. If you have a

Primer has been sprayed on this front wing and preparation carried out as described in the text. An aerosol can of black paint was then used to apply a guide coat. The guide coat only exists to be rubbed off! If you sand using a rubber sanding block and 240 grade paper, then low spots will remain with black paint, while high spots will show as shiny silver when you break through the primer to bare metal!

different coloured primer, choose an aerosol of a contrasting colour for the guide coat. Leave the guide coat to dry.

When the guide coat is dry, start to rub down the panel with 240 grade paper. Use the paper wet, with some soap in the water to prevent the paper clogging with paint removed from the panel.

If any low spots remain in the panel you should be able to pick them out after rubbing down with 240 grade on a rubber sanding block. If there is a low spot, the black will remain in a little patch, while all around you will have cut through the black paint to reveal grey or yellow primer.

Continue this process until you have rubbed down the entire panel. Make good any low spots with some more body filler if there is a major problem, or more Cellulose putty for minor problems. The idea is to have a perfect panel. Remember to pay attention to all edges, and corners. These can often be overlooked during preparation and any faults only come to light when you start spraying.

When this rubbing down is complete, clean off the panel with clean water, rub down with a clean dry cloth, then wipe down with panel wipe or spirit wipe.

Next apply another couple of coats of primer, depending on budget and how good a finish you require. When this primer has flashed off, again spray on a light dusting of black from an aerosol.

This time cut back with 600 grit paper used wet. When this has been done you should have a perfect panel without any high or low spots.

If your car is going to be used as an every day vehicle, then you can start to put on the final colour coats. If you want a "concours" finish, then continue the above process with 800 grit paper and finally 1200 grit paper.

Another view of the lower front valance on our Cortina respray project. We have applied plastic body filler to both ends of the valance and next we will apply filler to the centre section. When all the filler has been applied it will be flatted off, as described in the text.

VALANCE

If you are in any doubt about a panel, you can linnish it with a rough sanding disc, either by machine or by hand, then apply a thin wipe of body filler all over the panel. On our Cortina respray we used this procedure on the lower front valance. Although the panel was new, it had been welded onto the old panel along the sharp edge just below the grill. This welding resulted in a little bit of distortion so Terry decided that rather than spot-fill the valance we should wipe it with filler.

Mix up enough filler to do one half of the panel. Apply with the plastic wipe provided with body filler, and try to get a nice even coating of about one eight of an inch. When this has hardened, use either a body file or a sanding machine to even out the filler.

The aim is to have a very thin covering of filler all over the panel. This hides any low spots in the panel. If there are any high spots then they need to be identified and tapped down with a suitable hammer.

When the job is completed on one side of the panel, mix some more filler and continue on the other half of the panel. Blend in at the joining point and continue sanding.

When this is complete, follow the procedure discussed above using an aerosol as a guide coat for rubbing down.

Remember that the valance may not be seen in any great detail when the car is finally assembled. Much of it may be hidden by grill, bumpers, spot lamps and so on. You may even wish to spray the lower valance with some protective coating such as underseal. If this is the case, you need not spend a lot of time trying to achieve the perfect panel.

UNDERNEATH THE ARCHES

I know that this next bit is out of step, but if your restorations are anything like mine, they do not follow a nice tidy pattern! You

Problems on this Cortina boot lid. The filler has been rubbed down but in this case has been rubbed too much and is now "low" with regard to the surrounding steel. This is proved by the shiny area surrounding the filler. The solution is to apply more filler and rub down very carefully. Curving areas like this need more care.

may have prepared bodywork only to find that extra repairs are needed.

You may also have found some work that you did a year or so previously which you are not too happy with. You want to do it again.

The next section takes you through a major repair and welding job. Remember that the better the metalwork, the easier the subsequent preparation.

One of the most most common repair jobs on a Ford is welding a a patch on top of the MacPherson-strut tower.

The MacPherson strut is fitted to lots of cars but for years it was identified with Fords. When the MOT test was introduced it became common for Fords to be failed because of severe corrosion at the top of the strut mounting. The specialist dealers were quick to spot this and they produced a repair patch. This has to be welded.

Patching this area must be done with care. There is no point in putting a new repair panel on top of the already rusty steel. This is what happens all too often though. You ***CAN*** complete this repair so well that it cannot be detected. On the other hand there are some real disasters around! Remember is you want to show off your car at a rally, you are probably going to have the bonnet raised. Everyone will then judge your metalwork.

You may need to allow a weekend to replace a MacPherson strut tower, assuming you have had some metalworking experience before. If you have never tackled a job like this, do not start on a MacPherson strut tower. It is just a little bit too complicated for a complete novice, and you can really put yourself off if you make a hash of the job and have to call in a professional!

The correct way to go about it is like this;

1) Put some sort of protective blanket or dust cover over the engine. Then with a rotating wire brush get rid of all the loose

A very peculiar intrusion here, in a final bodywork preparation chapter, of a MacPherson strut welding repair. Was it done as a filler because he had little further to say on body-work? Guard against this sort of thing when writing articles or technical reports.

This stuff is brilliant at finding out how good your preparation really is. Spray some over your prepared surface (use a contrasting colour, say black over red), then flat down the aerosol paint. If you have high spots they will show up as shiny where you break through the paint, while low spots will not get flatted and will retain the aerosol paint. If all the "guide coat" gets rubbed off then your panel is perfectly prepared.

rust, paint and underseal used to disguise the area.

2) When you have done this, have a good poke round with that trusty screwdriver. If there is just some ragged metal left, but the thicker top of the strut tower is still good, you can go ahead and patch. However, if the strut tower is badly corroded you are going to have to consider some major surgery.

You will probably find that the vertical parts of the strut tower have not rusted much -- perhaps only the top inch or so, where they are welded to the upper pan.

3) Once the suspension unit and all the other mechanicals have been removed, get the hammer and chisel out and break the spot welds holding the verticals to the inner wing. Although we said it is probably in good condition, it makes a much neater job to replace the whole assembly.

4) Break all the spot welds and remove the verticals. The job is much easier if the wing is removed. However, you can still do the job if the wings are left in place. On some cars it might be a simple job just to unbolt a wing.

5) The round pan which forms the top of the tower will probably be badly rusted. It is welded to the verticals, to the top of the inner wing and to the side of the inner wing. It will take a fair bit of effort to release all the spot welds. The unhappy part comes when you remove the top and find that there is very little metal left in the inner wing. In some cases you have to replace a section of the inner wing. Your repair will form two thicknesses of metal. The first or lower thickness is the strut top, the second is the inner wing. You might also decide to plate the top with a proprietary repair patch. This will then mean three thicknesses of steel.

6) I like to assemble the tower top and verticals with clamps and bolts. The top can be bolted using the three bolts which hold

Spraying Hammerite

Hammerite is suitable for spray-gun application.

Thin 2 parts Hammerite to one part Hammerite Brush Cleaner and Thinners.

Apply 3-4 coats quickly to achieve a hammered metal finish.

Allow to touch dry between.

Shake gun before and during application.

It is recommended to use the spray-gun at 25-35 psi.

Unthinned Hammerite will cover approximately 4.5 square metres per litre, dependent upon absorbency of surface and method of application.

the MacPherson strut in place.

7) When all is in place, and the parts fit well together, tack-weld them.

8) The original line where the verticals were removed should still be visible. Now it is just a matter of welding the verticals into place, following the original line of spot welds.

9) You might need to dress the tower top so that its edge touches the inner wing. This is important to gain extra strength. Try to aim for about 1 to 2 inches or weld in this region. This is more than when original, but will do no harm.

10) When all the welding is done, treat the steel immediately with rust inhibitor or primer. A good heavy undercoat completes the job underneath.

Anyone with some metalworking experience should be able to do a repair to one side in a weekend. Once again do not depend on having the car ready in a few hours. This is when jobs start to go wrong and get bodged. Take your time and it will last another 25 years.

As I have already mentioned, I would not recommend you try this as a first job. It can really put you off when you get to the point where all the old strut is off and you have the daunting prospect of welding the new one on.

If the job is beyond you, there is no harm or shame in passing it to a professional. You can always try your hand at something a little easier.

Chassis Painting

Refer to the photographs if you want to paint your chassis with Hammerite paint. There is also a special section, above, describing how Hammerite can be sprayed.

Rustproofing*

After you have done all the restoration work, and your pride and joy has a sound chassis again, you will want to ensure that you do not have to light the welding torch again for a long time. There are many ways to protect the underside of a car, but basically they all involve preventing water from coming into contact with bare metal. The large flat areas of the floor pan will only give trouble if the carpets and underfelt get wet--in other words they will rust from the top down. Think about that. On a flat sheet of steel there is nowhere for water to gather and cause rust, it just drips off.

The main problem areas are the box sections which you have patiently repaired. This is where mud and debris can build up. This material stores water, acting like a sponge and this soon leads to rust. If you intend to rustproof the whole of the car it pays to have the underside washed with a high pressure washer or steam cleaner. Many garages now have these on the forecourt and you do the work yourself. Make sure you get all the mud dirt and loose underseal off. Put the jet into all the awkward areas, like under the wheel arches.

When you have done the cleaning, get some

* Major rustproofing is often better left until bodywork repairs and spraying have been completed.

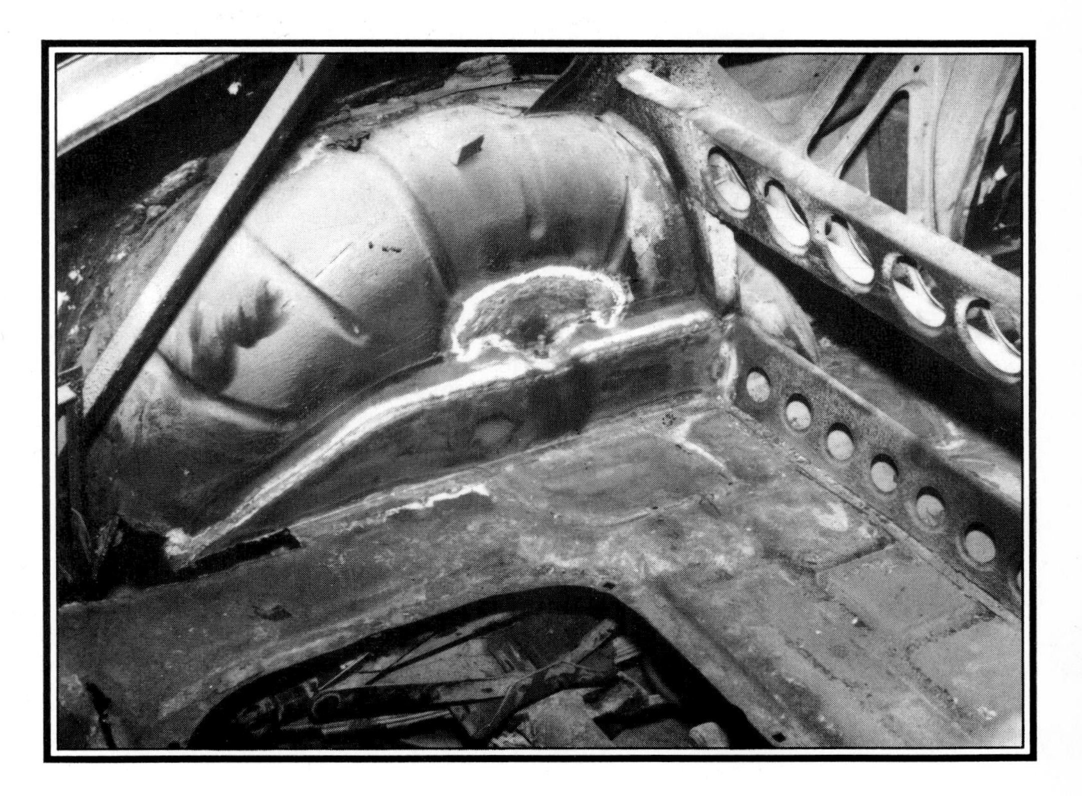

Repairs, such as the welding being done on the inner wheel arch and chassis leg of this Lancia Fulvia Sports, need to be primed as quickly as possible. Also remember that dirt and dust surrounding the repair needs to be removed prior to spraying.

rags and mop up as much of the loose water as you can from the underside of the car. Some people take it for a short run to blow away excess water. Make sure its a dry day though! Dry off each area before applying any rust fixers.

It is difficult to give general instructions about rust inhibitors. You will have to follow the instructions which come with the product. Generally they have quite clear instructions.

Basically what you have to do is clean off any loose paint, dirt or other debris. Paint on the rust inhibitor as described by the manufacturer. Sometimes you have to apply a second coat, while other have to be washed off with water. It pays to treat a small area at a time.

Follow the instructions closely. When the rust-proofer has cured for the required time, you can paint on underseal or bitumen.

Underseal

There are several kinds of underseal. Again, you have to remove all loose paint, dirt or old underseal so that the new underseal will bond tightly to the body. If you want to put on a second coat, do so, but allow the time for the first coat to dry. This will be marked on the tin.

Do not get underseal on the exhaust system. If you do you can wipe it off with a cloth soaked in White Spirit, Paraffin or Turpentine. The same solvents can be used to wipe any underseal from paintwork where it will show.

Now that the outside is rust inhibited and coated with a nice thick layer of underseal what can you do to protect the insides of the box sections?

Manufacturers drill holes in the box sections to allow water and debris to drain out, but over the years these block up and the rust starts. The holes also allow the boxes to

Finnigans Brush on Metal Primers

A metal primer which stabilises rust and prevents rust creep is ideal for coating repaired areas of a vehicle body by brush. Finnigans No. 1 Metal Primer is ideal for this purpose. (Thanks to Hammerite Products Ltd for the following information).

Hammerite Products Ltd recommend that a simple compatibility test be carried out if large areas are to be coated, as rectification could be costly.

Finnigans No. 1 Metal Primer is a rust stabilising primer for application to prepared rusty surfaces. No. 1 is not acid based but works by penetrating and stabilising existing rust. This process prevents rust creep as well as providing a smooth primed surface. Finnigans No. 1 is lead free and can eliminate expensive surface preparation whilst providing a stable surface ready to paint.

It can be used as a primer under a wide variety of top coats to improve performance and protection.

It will withstand temperatures of 300 degrees C (maximum).

Two coats are recommended. It is touch dry in approximately 1 hour at 20 degrees C. It can be overcoated with alkyds and itself after 4 hours. It can be overcoated with epoxies or acrylics after 6-8 hours.

Remove all loose or flaking rust, scale old paint and any other surface contaminant by wire brushing, scraping or other appropriate means. Any oil or grease should be removed using an appropriate solvent. No. 1 can be applied to surfaces prepared to a lesser standard but may not perform as well.

Apply by brush, roller or conventional or airless spray. Two coats are preferred for maximum rust stabilisation.

For conventional spraying thin with Hammerite Thinners at 10 to 15 per cent. Set air pressure to 25-35 psi with the nozzle set to maximum.

No. 1 can be removed from unwanted surfaces by wiping off with Hammerite Thinners.

Finnigans No. 1 Metal Primer is not a holding primer and must be overcoated with a suitable top coat for long term protection.

Suitable top coats include: Conventional oil based and alkyd paints, Epoxies, Polyurethanes, most Cellulose types, Acrylics, Bitumen paints, most water bourne coatings.

Finnigans No. 1 Metal Primer Health and Safety Precautions

No. 1 Metal Primer is FLAMMABLE, with a flashpoint of approximately 32 degrees C.

In case of contact with eyes, rinse immediately with water.

Remove splashes from skin with soap and water or a recognised skin cleanser.

Keep away from sources of ignition. (No smoking).

Contains no added lead.

Keep out of reach of children.

For further information, contact the Technical Services Department of Hammerite Products Ltd. (See Addresses section at the end of the book).

ABOVE: The MacPherson strut towers and top pan clamped in position, ready for a trial fitting of the top plate.

BELOW: The top plate welded into position ready for an edge to support the wing.

The same MacPherson strut repair as shown on the previous page. All the welds have been ground down with an angle grinder, primer has been sprayed on and any further smoothing or dressing has been done. The dusty appearance on the inner wing is due to the primer having been rubbed down "wet".

breathe, thus cutting down on damaging condensation.

These holes can be used to spray a rust inhibitor inside the box section. For this purpose, companies like Waxoyle provide an applicator tool which is small enough to go into these holes.

Get as much muck out of the box sections as you can. Poke it out with a welding rod or screwdriver, or if you have the facilities blow the muck out with compressed air. You must wear goggles if you do this, and keep everyone else clear.

If you have been working on a box section you should have cleaned out as much muck as you could while the box was open. If you need to get in from above, then look for holes plugged with grommets, or drill your own holes. Once you have applied the rust treatment, plug the hole with a grommet. You will need about 5 litres to treat a small car, possibly double that for a bigger saloon such as a Hillman Hunter or Ford Cortina.

When you have completed the rust proofing process and you are sure that no water is entering to form puddles on the floor, you can lay some extra underfelt to reduce noise levels.

When the car is ready...

Assuming that you continue the preparation until the vehicle is 100% ready to be sprayed, the next step is to prepare the spray booth and your equipment.

Preparing the spray booth

If you are going to spray in your garage, you need to prepare the garage. The biggest enemy of a spray painter is dust, so get out the vacuum cleaner and special tools and vacuum all the shelves, beams and rafters to get rid of as much accumulated dust as possible.

This bonnet is having a very light covering of plastic body filler applied all over to smooth out the many wrinkles and minor dents. You don't do this on every panel. Most panels will only need filler in small areas.

If necessary, use the air line from your compressor to blow into difficult corners. This will soon shift the dust. Better to find it now than later, on your newly sprayed panel!

This cleaning is going to take a couple of hours, so get help if you can. Remember that you will not be able to reach all the corners with a vacuum tube, so you will need a selection of household brushes and dust pans. You should aim to fill a couple of dust bags if you do this job properly!

Once all the dust is out of the way, tidy away as many things as possible from the surrounding area. Once you start to spray you don't want to start tripping over bits of engine and so on.

Put things away in cupboards, or move them to another storage area. The more room you have to work the easier the job is going to be.

If you have the facilities, you can drape plastic sheets all round the workshop. These will help to keep any remaining dust at bay, especially dust lurking in corners or on the brickwork of the walls.

Next, sweep the floor. Once you have finished, sweep it again. You will be amazed how much dust you pick up on the second sweeping! If you are still in doubt about how clean the floor is, sweep it again. The reason for all this cleanliness is that as you walk around the floor with a compressed air spray

This is the photo you all wanted to see!! Technically known as "runs", "sags" or "curtains" they were caused by too thin a mixture. Tommy thinned his primer twice (by accident) and this is the result. Leave to dry, rub down to flatten the runs and spray again -- this time with correctly thinned paint..

in your hand, all manner of dust will be moved about in the air and will fall on your nice shiny, wet paint. You cannot take enough trouble to get the garage clean and dust free.

When you come to paint, many authorities recommend that you sprinkle a layer of water on the floor to damp down any remaining dust which may be present. This seems like good advice. Don't overdo this process however, or you you may run the risk of splashing the new paint with water droplets instead of dust.

Remember in a traditional British brick garage you are going to be short of room. If you have a small car then you will be able to get all round it, but anything larger may mean having to move the car in the middle of the job.

How do you do this? If you only have access to one side of the car, then carry out as much spraying as you can. Leave the spray coat to dry for the recommended period, then get some helpers to push the car out, turn it round and push it back in again. A few words of warning. Earlier in the book we said that once a metal surface was prepared for painting it should not then be touched by hand. Make sure your helpers are wearing cotton gloves when they push the car.

The above situation is not ideal, but you may have no choice. Remember that if the wheels are masked with plastic bags, then you must remove these before the car can be moved. Also, once the car is back in the garage again you will have to ensure that the tyres are clean and that no more dust has been brought into the workshop. Mask the wheels again and check for dust before resuming spraying.

Workshop Temperature

Ideal painting temperatures range from 5°C to 30°C depending on the conditions in the workshop and what paints are being used.

For Cellulose, Two-Pack and Synthetics the ideal temperatures are approximately 20°C, that is, room temperature.

Anything between 5° and 20°C will be all right but on warm summer days temperatures are obviously higher. Higher temperatures can be accommodated by adjusting with slow thinners or hardeners.

Above 30°C is a stoving temperature and requires a professional spray booth. A spray booth will warm up to about 80°C. Fifteen minutes at 80°C is ideal for Two-Pack.

Never stove Cellulose as it is an air drying paint. Similarly, don't stove Synthetics unless a Two-Pack hardener has been added to the synthetic.

Full Respray?

If you are spraying a complete bodyshell the chances are that you changing the colour. Even if you are not changing the colour the following section will tell you all you need to know about the sequence of events in spraying a complete shell.

First spray the door jambs. This is the area round the A post, B post and C posts. Set the spray gun to the small round spray pattern shown on page 106. You should also reduce the air pressure to limit the amount of overspray created. This is particularly important if the shell is not fully stripped out, that is, there is upholstery and trim panels installed.

When you are happy with the finish on the door pillars, next spray *inside* the boot lid, bonnet, and along the bottoms of the front wings and front and rear valances.

When this work is complete, follow the guidance given on page 77 which shows one possible sequence of spraying a complete bodyshell.

There are other ways to work round a shell, but the key to success is to work *away* from the panel you have just sprayed. Refer to the diagram on page 77.

This product, Naval Jelly, removes rust from all metal surfaces. Brush it on, following the manufacturer's instructions on the rear of the bottle, then wash it off.

Chapter Four

Tak Rags from Gramos. These specially impregnated rags are used to wipe over a panel just before spraying. The rag is slightly sticky and picks up any dust or fluff still on the panel. These Tak Rags cost about 50 pence each and are well worth the money. Available from your paint supplier. Get some.

Pre-Painting Checklist

1) Is preparation of the body complete? (There is no point finding out after you spray a panel. Wipe a wet sponge over your panel and look for ripples, lumps or other signs of poor preparation). If you find any, put them right, now!

2) Are the panels clean? Wipe down panels with pre-cleaning or degreasing solution.

3) Are your air hoses clean?

4) Is your spray-gun clean?

5) Is the gun adjusted properly and does it have the correct set-up for the material being sprayed?

6) Is your paint thinned correctly, and is it mixed thoroughly?

7) Is the vehicle masked carefully and correctly?

8) Are there any naked flames or sources of ignition present?

9) Is the spray booth temperature correct?

10) Do you have enough material to complete the job?

Spraying Outside

You can spray a vehicle out of doors in the UK but you must be very careful about the weather conditions. If the day is dull and overcast there may be a risk of a lot of moisture in the air, so a warm dry day is a better choice.

You do not want any wind either, otherwise all sorts of airborne nasties will be blown onto your new paint. Even on a still day you run the risk of insects landing on the new paint.

Overseas readers may be in a better position to spray outdoors.

Remember that all the solvents will be

Customising offers the spray painter unlimited opportunities to try different colour schemes and two-tone effects as on this smart and very tidy Volkswagon Beetle. Known as the Blue Iguana it was on show at Tredegar House, Newport, Gwent in September 1993.

The petrol tank is black, so preparation must be absolutely perfect otherwise any little blemish will show through. The coach lines around the AJS badge are gold and painted by hand -- a very steady and skilled hand. Although these old skills are dying out, we can still admire the craftsmanship from the 1950s in these old motorcycles.

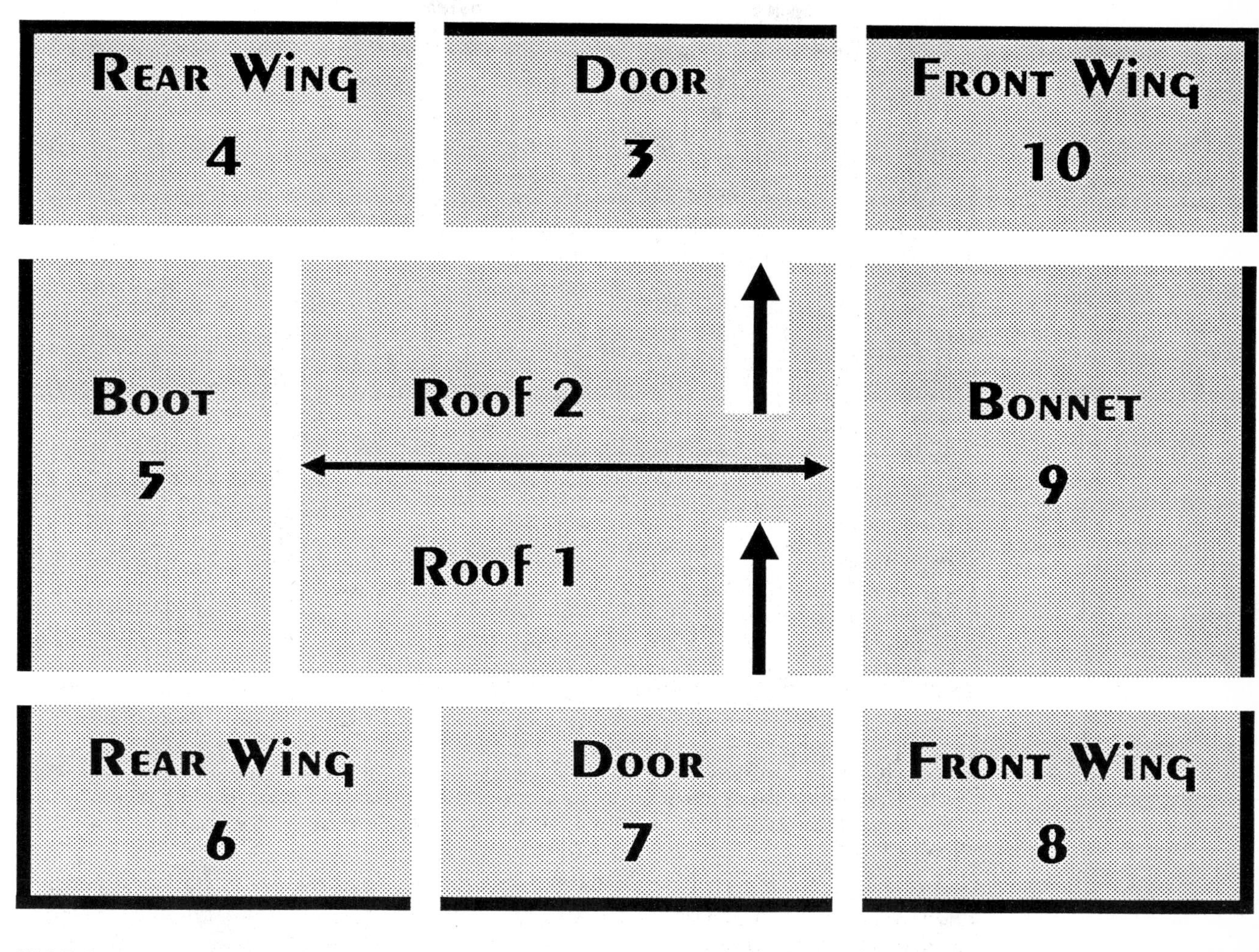

WORK OUT YOUR SPRAYING SEQUENCE IN ADVANCE.

The following guidelines are offered for newcomers to paint spraying. More experienced sprayers (such as Trevor) may tackle the job in a different way. The end result is the one that counts!

Generally, if you are spraying a complete, stripped-out bodyshell you will do the following panels first: Insides of sills, boot lid and bonnet, insides of doors and door pillars. Insides of "A" posts, "B" posts and "C" posts as required. When the inside of the shell is complete, the spraying sequence is generally as shown above.

ROOF: The roof panel is usually split into two or more sections. Problems may occur when spraying the roof as the gun may have to be tilted and paint may drip onto the job.

Start from the gutter and work away from yourself, towards the centre of the car. When you reach the half-way point, move round to the other side of the roof and work FROM the wet edge towards the gutter. Ensure a "wet edge" is achieved as you proceed round the roof panel.

DOORS: On a two door car, spray the door on the same side as the front wing just completed. On a four door car complete the front door, then the rear door.

REAR WING: When the door(s) are completed, spray the rear wing or rear quarter panel.

BOOT: The boot lid, and area surrounding the boot are tackled next. Rear valances can be sprayed at this stage.

REAR WING: Continue working round the car, tackling the unpainted rear wing or rear quarter panel next.

DOORS: Spray the rear door, followed by the front door on a four door car.

FRONT WING: Complete the sequence round the car by spraying the unpainted front wing next.

BONNET: Most experts recommend that the bonnet is painted last. This is because any overspray from previous panels can be corrected at this stage. However, Trevor sprays the front wing last! (The bonnet and front wings are usually considered to be the "make or break" panels on a car, where the entire paint job is judged).

FRONT WING: After the bonnet is complete, tackle the front wing on the side of the car where you COMPLETED the roof panel.

Second Coat: When Trevor sprays a second coat he works round the car in the opposite direction to the first coat. This method ensures that the dry edge (the last edge to be sprayed) is on a different panel. This way, a dry edge does not build up on the same panel after each coat.

No apologies for another photo of KPU 381 C. This ex-works rally car was restored from "wreckage" (owner Bryan Moorcroft's own words) to the immaculate state you see above. To the right is another example of a two-tone vehicle where the join between the colours is disguised by chrome strips.

released into the air so you may have extra problems if there are fish ponds, pets or children nearby. Other vehicles will not like your overspray either!

If possible, try to spray in a garage or proper spray booth.

Air Line Lore

Remember that as compressed air passes through a tube it loses pressure. This is particularly true the longer the length of air line used, and the smaller the diameter of the hose. Refer to the Reference Chapter at the end of the book for details of air pressure drop.

Another thing to remember about air hoses is that the outsides of them get dirty.

It will pay you to wipe down your air hose with a damp cloth. This will remove any dust and dirt which is sure to appear on your nice shiny new paint.

Chapter Two has some information about compressor care. The main things to remember are to drain the holding tank after each spraying session and to follow the manufacturer's instructions regarding maintenance.

Now We Spray!

Having shown great patience to get this far without trying to use the spray-gun, it is almost time to apply paint!

The next chapter takes you through many stages, including basic spraying technique, basic faults in spraying technique and some hints and tips for first-timers. There is also a section on paint related faults.

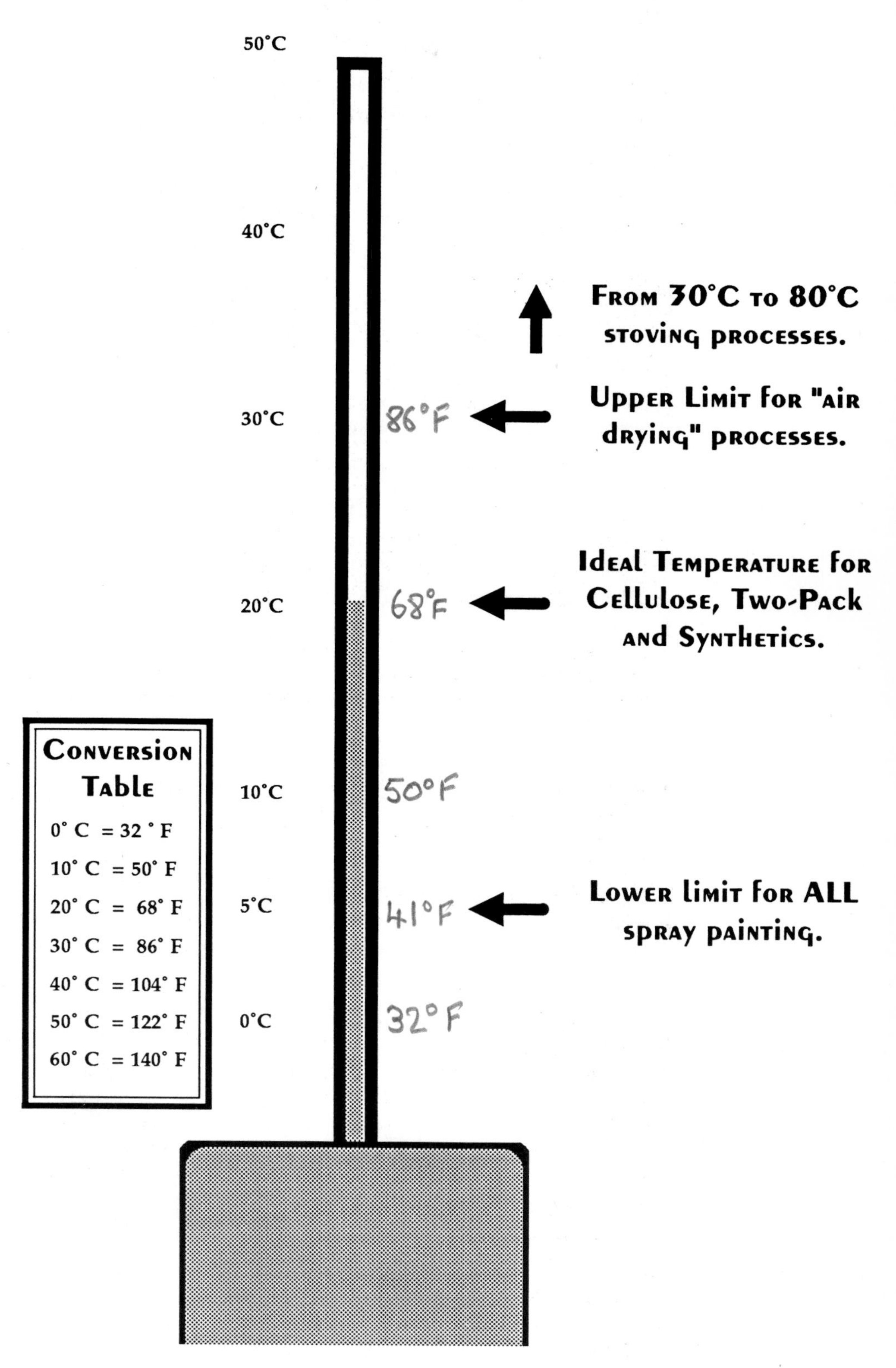

Conversion Table

°C	°F
0° C	= 32 ° F
10° C	= 50° F
20° C	= 68° F
30° C	= 86° F
40° C	= 104° F
50° C	= 122° F
60° C	= 140° F

Preparing Rusty Surfaces

Where rust has occurred on a body panel, treatment is required to contain the rust, otherwise it will eventually break through a newly painted surface. Hammerite Products Ltd recommend their product Kurust for this purpose. (Thanks to Hammerite Products Ltd for the following information).

Kurust

Kurust is a water reducible polymer based rust converting primer for prepared rusty surfaces. Iron oxides are quickly converted into stable and insoluble and mettallo-organic complexes which makes a suitable base for the application of paint systems.

Only one coat is recommended.

Kurust is touch dry in 15-30 minutes at 20 degrees C. and ready for over coating after 3 to 4 hours depending on temperature, humidity, air movement etc.

Remove all loose or flaking rust, scale, old paint and any other surface contamination by wire brushing, scraping or other appropriate means.

Any oil or grease must be removed using an appropriate solvent.

Kurust can be applied to surfaces prepared to a lesser standard but may not perform as well.

Kurust can also be applied to blast cleaned steel to prevent flash rusting.

Apply preferably by brush or roller. Conventional spraying is also acceptable.

Avoid applying excessive amounts of Kurust leading to runs or pools of unreacted polymer.

Apply within the temperature range 5 degrees C to 40 degrees C. and with a maximum humidity of 85%.

Thinning is not recommended.

Use water for cleaning before Kurust dries, or Finnigans Brush Cleaner and Thinner once the polymer has dried.

Kurust is not a holding primer and must be protected with a suitable paint system, preferably within 48 hours of application.

Kurust is compatible with most finishing systems, including: Alkyd, Chlorinated and acrylated rubber, Epoxies, Urethanes, Vinyls, Nitrocellulose and acrylics.

Kurust Health and Safety Precautions

Kurust has a flashpoint above 100 degrees C.

Kurust is acidic and harmful if swallowed.

Avoid contact with skin and eyes.

In case of contact with eyes, rinse immediately with water.

Remove splashes from skin with soap and water or a recognised skin cleanser.

Contact with skin may cause staining.

For further information, contact the Technical Services Department of Hammerite Products Ltd. (See Addresses section at the end of the book).

This beautiful 1952 MG YB sports saloon is believed to have won the 1953 Welsh Hill Rally. Bought in 1970 by Sheila Taylor the car was fully restored in 1986 by Mike Phillips of MGP Restorations, The Old Telephone Exchange, Brassknocker Street, Magor, Gwent, NP6 3EG, phone number 0633 881199. Multiple coats of Two-Pack were used to achieve a deep and lasting gloss. The car is used almost every day, and as Sheila says, hardly ever gets polished!

Protecting the Underbody

For underbody protection a good underseal is required. Finnigans Underbody Seal is excellent for this purpose. (Thanks to Hammerite Products Ltd for the following information).

Finnigans Underbody Seal

Finnigans Underbody Seal is a highly thixotropic bitumen modified wax based corrosion preventative. It has a good long term flexibility, abrasion resistance and excellent corrosion prevention properties.

It provides abrasion resistance and anti-chip properties on exposed surfaces of vehicles, for example wheel arches, chassis and sills. It can be applied over Finnigans Waxoyl.

Underbody never dries to a hard tack free coating but offers much better abrasion and anti-chip resistance than Waxoyl, especially in thicker films.

Remove heavy rust deposits and heavy scale using power tools, hand tools or high pressure water washing.

Surfaces may be damp but not running with water. Minor oil or grease deposits should not affect the performance of Underbody but for best results should be removed.

Underbody can be applied by brush or roller. Up to 1.5mm wet film thickness can be applied without sagging in one coat.

A spray viscosity grade of Underbody is available. For further details contact the Technical Services Department of Hammerite Products Ltd.

Underbody can cleaned off surfaces with White Spirit.

Finnigans Underbody Seal Health and Safety Precautions

Finnigans Underbody Seal is FLAMMABLE, with a flashpoint of approximately 43 degrees C.

In case of contact with eyes, rinse immediately with water.

Remove splashes from skin with soap and water or a recognised skin cleanser.

Keep away from sources of ignition. (No smoking).

Do not breath vapour/spray mist.

In case of insufficient ventilation wear suitable respiratory equipment.

Keep out of reach of children.

For further information, contact the Technical Services Department of Hammerite Products Ltd. (See Addresses section at the end of the book).

A fine example of the 1951 Riley RMA 1.5 litre, exhibited at Tredegar House, Newport in September 1993. Again, the two-tone paintwork has a chrome strip over the join between the white and maroon. Notice that on this model the hub caps are painted white but with chrome inner and outer borders.

Two Lotus Cortinas. Both had the green stripe sprayed after the white. But look at the green stripe above the bumper. Which is correct? You had better find out BEFORE you mask your Lotus Cortina. It is a tiny detail but important if the car is being restored to its original condition.

Protecting box sections

For protection of box sections, such as 1960s monocoque bodyshells, Finnigans Waxoyl is recommended. (Thanks to Hammerite Products Ltd, for the following information).

Finnigans Waxoyl

Finnigans Waxoyl is a corrosion inhibiting coating combining excellent water displacement properties, long term flexibility and good crevice penetration properties with good rust suppression and prevention properties.

Waxoyl need one or two coats, depending on service conditions.

It can withstand up to 150 degrees C, and never dries to a hard abrasion resistant coating. It should NOT be used where abrasion resistance is required.

Remove heavy rust deposits and heavy scale using power tools and or hand tools.

Surfaces may be slightly damp; minor oil and grease deposits should not effect the performance of Waxoyl but for the best performance should be removed.

Waxoyl can be applied by brush, roller or spray equipment. It may require more than one coat to achieve correct dry film thickness if applied by brush or roller.

Waxoyl is primarily designed for use with Finnigans hand pumps, pressure units and extension probes. Airless spray equipment, however, is a satisfactory method of application.

Waxoyl can be cleaned off surfaces with White Spirits.

Waxoyl cannot be overcoated with any other coating, apart from itself or similar wax-based coatings.

Waxoyl Health and Safety Precautions

Waxoyl is FLAMMABLE, with a flashpoint of approximately 42 degrees C.

In case of contact with eyes, rinse immediately with water.

Remove splashes from skin with soap and water or a recognised skin cleanser.

Keep away from sources of ignition. (No smoking).

Do not breath vapour/spray mist.

In case of insufficient ventilation wear suitable respiratory equipment.

Keep out of reach of children.

For further information, contact the Technical Services Department of Hammerite Products Ltd. (See Addresses section at the end of the book).

YVP 464

Above, and opposite page. Mike Phillips exhibits his hand-built classic, named the SILURIAN. Based on a 1979 3-litre Jaguar, all the aluminium coachwork was hand formed and took three years of part time effort. An etch primer was used followed by a high fill primer. Two-Pack primers were used and about 8 coats applied. Each coat was painstakingly rubbed down before the next was applied. When all the priming coats were to Mike's satisfaction, some 15 coats of Cellulose were sprayed. Having done all the development, Mike would be willing to build replicas. Contact Mike at MGP Restorations, The Old Telephone Exchange, Brassknocker Street, Magor, NP6 3EG. Telephone 0633 881199.

No book on paint spraying would be complete without a photograph showing the correct distance between the gun and the panel being sprayed. Between six and nine inches is about right for most jobs. See Chapter Five.

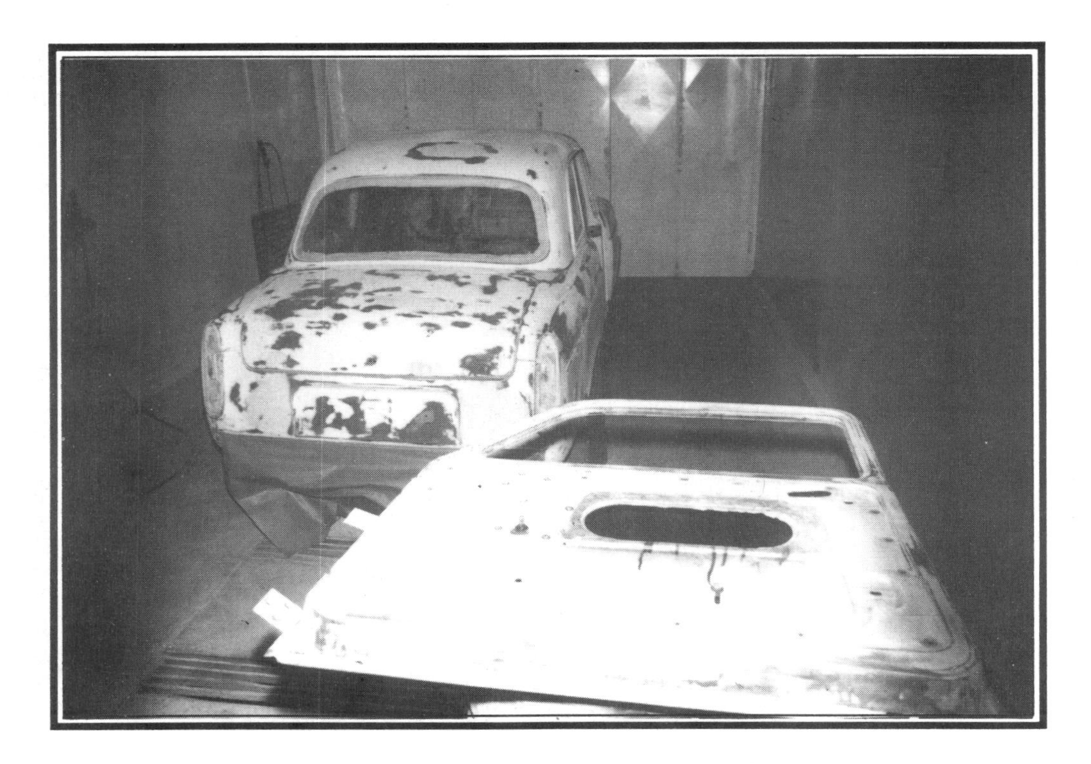

ABOVE: Bert Wiesfeld's 100E Ford is being masked prior to spraying. The door in the forefront of the picture will be sprayed off the car. Looks like Bert is lucky enough to have a real spray booth available. This means he has room to move around the car when spraying. Photo: Bert Wiesfeld.

BELOW: Over the epoxy primer goes a coating of filler/primer. The sprayer has sprayed the door with primer. The sprayer is using a gravity-fed gun which is often used on the Continent. Photo: Bert Wiesfeld.

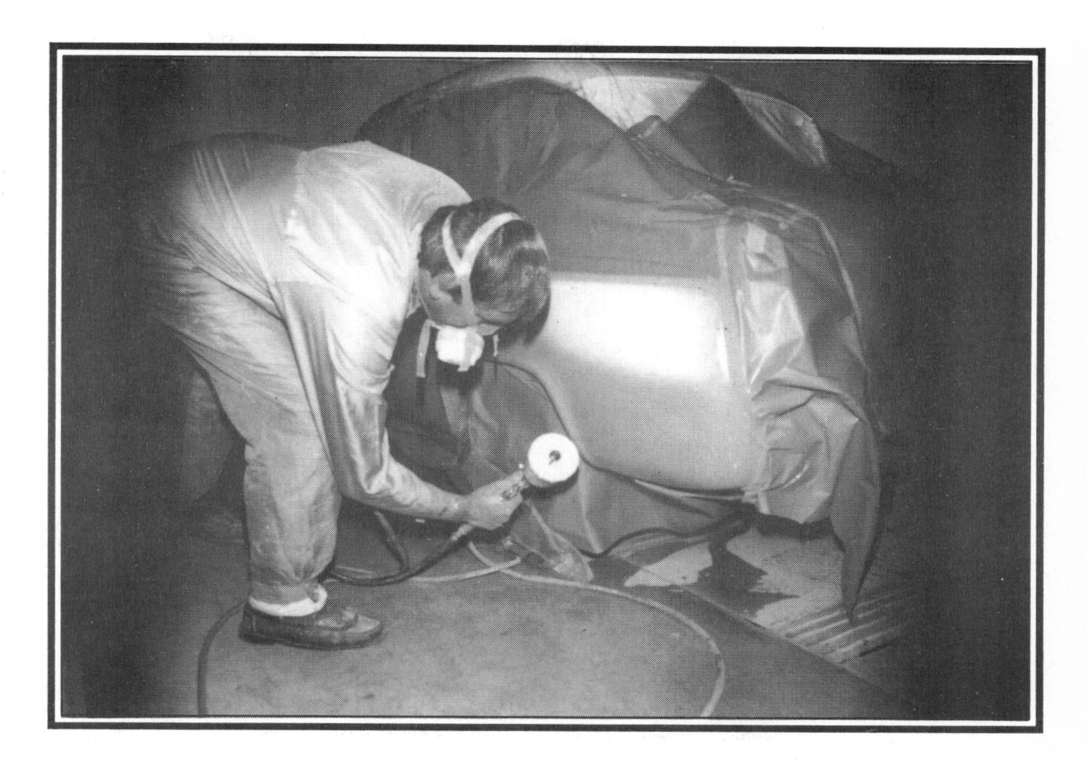

ABOVE: The car has been sprayed in white and masked off again. Here the sprayer is applying the brown for the lower half of the Ford 100E. Photo: Bert Wiesfeld.

BELOW: The complete car showing off the beautiful paintwork. As we have seen many times in Paint Craft, where the two colours meet the designer added a chrome strip to cover the join. Photo: Bert Wiesfeld.

Basic Technique & Faults

Cleaning and De-Greasing

Prior to masking anything, your preparation must be complete. Check that all existing paint has been sanded down properly, and that any local repairs have been featheredged.

If you are in any doubt about what the finished job will look like, wet a sponge and wipe it over the area in question. The wet film of water will show up any problems. Those same problem areas will shine through the newly applied paint too!

When you are 110% satisfied with the preparation, take some steps to clean the area to be sprayed. By this we mean de-greasers.

In the good old days, the home sprayer wiped his panel down with petrol to remove dirt and grease. Nowadays we can do a lot better. There are several de-greasers on the market and they are well worth the money. For the few pounds spent on a tin of panel wipe, you are helping to avoid failure with perhaps hundreds of pounds worth of paint.

Wipe the panel with the de-grease solution and a CLEAN non-fluffy rag. Using another clean non-fluffy rag, wipe the de-greaser off the panel. The two rag method speeds up the drying process. Follow any instructions on the de-greaser can. If there is a waiting period specified for the de-greaser to evaporate, don't rush on to the next step. Wait until the specified time has expired.

Masking

Masking consists of covering over items which must not be spray painted. This usually involves wrapping the items in sticky tape, or brown paper, or both.

Don't fall into the trap of masking the car THEN start to prepare the surfaces. It must be done the other way round. If you need to protect a part, that's different. Protect with brown paper and or tape, but be sure to remove this protection before you mask up for spraying. Why? Because that protecting paper and tape will house all sorts of dirt, old paint etc. all caused by you preparing the other surfaces. Don't get caught. Keep things clean.

Masking is an art in itself

Masking Materials

Masking materials consist of brown paper, special masking sticky tape in various widths, and plastic bags. Each item has many uses as we shall see.

Brown paper is recommended because of its strength. It has one other major advantage over newspapers. In some circumstances the ink in newspapers can be washed out by the solvent being used. This running ink can get on the panel and spoil the entire job. Newspapers can be used, indeed most sprayers actually use them, but be aware of the possible problem. Use several sheets of newspaper instead of a single sheet. When it gets wet it will tear.

Sticky masking tape has several features you might not be aware of. It must stick to paint, glass, chrome, stainless steel, aluminium and rubber. When it is removed the glue must not contaminate the paint. And finally it must be strong enough to be pulled, twisted and stuck round a variety of shapes.

Tape comes in rolls and is available in various widths. You will probably need rolls 1 inch wide and 2 or 2-1/2 inch wide. Several rolls will be needed for the average car.

Masking - General Tips

Don't automatically think of wrapping masking tape round everything. This is wasteful in materials, not particularly quick and a devil of a job to remove when the spraying is complete.

Although the following section is designed to help you master the art of masking don't worry if you find a different way to mask something.

Some careful masking was required to get the black wing top line and the side stripe which is green. You'll need lots of brown paper, masking tape and a sharp pair of scissors and a craft knife. You may have some ideas of your own for masking.

Masking Door Handles

If you can, remove door handles prior to spraying. If not try this:

Instead of wrapping tape round and round D-shaped door handles, try masking from one side of the "D" to the other. Apply a piece of tape to the top, so that the edge of the tape is parallel to the body but not touching the door panel.

Continue to apply pieces of tape on the top and bottom of the door handle until it is covered.

Ensure that the very edges of the door handles are covered by tape, but also that the door panel has not been masked off accidentally. If you spray the door and find that a little bit of the door panel has been masked off by tape you have a fiddly job on your hands. You either respray the affected area or touch it in with an air brush if one is available.

Masking Door Locks

Generally locks can be masked with a small piece of tape wrapped right round the lock, then stuck down on itself to form a seal.

Another method is to stick a piece of masking tape over the lock, run you finger and thumb right round the lock to ensure the tape is firmly in place, then trim round with a very sharp craft knife. You must cut the tape, not tear it, so a sharp knife is essential.

Masking Glass

Nothing makes a respray look cheap and nasty as much as overspray on glass or rubber surrounds.

Mask the lower part of the glass first. This way if you have to overlap your paper anything running down the paper will run outside the paper and not find a path underneath the paper. The same principle is used on roofing tiles.

Here the wheels have been masked with brown paper, the door aperture has been masked from the inside so that the A-posts and B-posts can be sprayed. There is no glass fitted so the window apertures have been sealed with polythene and masking tape. The sprayer is using a gravity-fed gun to apply epoxy (Two-Pack) primer. Note that the operator is holding the air hose in his left hand so that it is kept under control at all times. Photo: Bert Wiesfeld.

Attach masking tape to one edge of your masking paper. Carefully apply that taped edge to the bottom of the glass area. If you are masking the rubber surround separately use extra tape.

Be careful to get edges right. One little mistake will either result in an exposed area of glass or rubber surround being sprayed OR an area of panel may be masked where it should be sprayed.

Where paper is overlapped, make sure you seal both edges over the entire length with masking tape.

Masking a Door Mirror

The quick and easy answer is remove the door mirror. If this is not possible, wrap a plastic bag round the mirror and secure with masking tape. Next, carefully mask round the base of the mirror where it meets the door. Yet again, careful attention to this sort of detail pays dividends when the masking is removed. Get it right and there is little or no corrective work required.

Masking Weatherstrip

A single strip of masking tape is the answer to masking weatherstrip. Ensure accuracy when positioning the tape. If necessary use several lengths of tape instead of one large length. Smaller pieces are easier to handle and don't stick to things they are not supposed to stick to.

Masking Gutters or Mouldings

As with weatherstrip, gutters respond well to a single length of masking tape. Follow the advice given for "Masking Weatherstrip" above.

Masking Upholstery

Upholstery varies so much that there is no definitive piece of advice we can give.

Large white dust sheets, of the type often used to protect furniture, are ideal for upholstery. Lay some newspapers over the seat panels prior to covering with a dust sheet. Ensure that the edges of the dust sheet are tucked securely underneath the seat or otherwise fastened securely. Nothing moves masking so quick as compressed air at 50 psi from a spray gun.

Masking Trim Panels

Door trim panels can be masked with paper and masking tape, but for a little extra work why not remove them? Remember if you do remove the interior door trim and then spray the insides of the door that you run the risk of spraying directly onto some of the glass. Take care to prevent this by stuffing paper or a clean cloth up inside the door to prevent spray going onto the glass.

Most door trim panels are held in place by some form of press stud. These studs are usually located about half an inch in from the actual edge of the panel. This little edge allows you to secure masking paper under the trim, then fold it over to cover the trim.

Masking Windscreen Glass

This is similar to "Masking Glass" above. The additional problem is that there may be a more complicated rubber surround which will require careful masking. On some vehicles the rubber surround may have a chrome or stainless steel decoration which also has to be masked.

As with glass work from the bottom upwards so that any over laps work to your advantage.

That rubber surround is generally masked and paint is then sprayed up to the rubber. It is a much better solution if you can either remove the rubber surround or lift the rubber to allow some paint to go underneath.

There are several tools available to help you do this, but you can simply use some plastic twine inserted between the rubber and the panel. When this is positioned correctly it will hold the lip of the rubber away from the panel and allow spraying to reach under the rubber. The end result is a much neater and more professional job.

Masking Lights

For round headlamps try this approach:

Cut a piece of masking paper about a foot long. Attach masking tape to the long edge, allowing half the masking tape to stick to the paper. Now stick the free edge of the tape round the glass of the headlamp, or round the chrome of the headlamp if this is still fitted.

As you work round the lamp, the paper and tape will buckle, and start to form a cone. You may need a second or even a third piece of paper and masking tape. Work round until you have a cone which projects outwards from the headlamp. Now simply join up the open ends of the cone with masking tape.

Treat round, double headlamps as two individual masking jobs.

For square lights the problem is simple. Select a piece of masking paper which will cover the light. Stick some masking tape to the top edge of the paper, again with about half of the tape stuck to the paper. Stick the free edge of the masking tape onto the top edge of the lamp. Fold the paper down over the remainder of the lamp then fix the remaining three edges with masking tape.

For small sidelamps, or rear lamps you can follow the procedure for round headlamps if the lamp cluster is large, or individually wrap tape round smaller lamps. Again think of the removal process before jumping in and masking without some thought.

For spot lamps or fog lamps follow whichever of the above methods suits you best. The secret is in the result!

Masking Radio Aerials

A simple thing like a radio aerial can cause the most clever mind to go blank. I bet you thought about wrapping masking tape round and round the aerial in a taper from bottom to top? Think about how you are going to remove this sticky problem when the spray job is done?

Consider this for a moment. Does the aerial

Masking an aerial need not be a difficult task. Don't wrap tape round and round! Mask the aerial with a single strip of tape as shown above. Another piece of tape may be required to mask off the base of the aerial. This method is quick, effective, and above all is easy to remove!

retract? If so your masking problem is considerably reduced. If it does not retract, have you a paper tube which would go over the aerial?

Failing that, how about taking a broad piece of masking tape, cut to the required length, and sticking it edge to edge to form a tube. Fit this over the aerial and adjust the sticky edges as required.

Or take two pieces of tape, place them on opposite sides of the aerial and press the edges together to form a tube.

When you come to remove it, a simple cut with a craft knife up the length of the masking take will remove it.

Miscellaneous Masking

Exhaust pipes which exit from the rear or the side of a vehicle may also need to be masked. Consider simply wrapping the exposed piece of pipe with masking tape.

You cannot win with badges. They are best removed. However, they are very hard to remove without damaging them. They are notoriously difficult to mask properly and if not removed will have the original paintwork underneath. This may not matter, but if a badge is stolen or damaged the old paintwork now exposed by the missing badge makes the resprayed vehicle a bit disappointing.

Where the badge consists of ornamental writing such as "Cortina" either remove it, looking on the INSIDE of the panel to see how it is fixed, or mask it with many small pieces of tape. This is tedious and time-consuming but worth it in the long run. Many of these badges are very difficult to replace now.

Masking Bumpers

Chrome bumpers and their associated brackets present a problem. The preferred choice is to remove them, but if you have ever struggled with rust nuts and bolts on a

A masking problem. How can you mask that "De Luxe 1500" badge under the reversing lamp? The answer is, not very well. It would be better if you could remove it. Similarly, remove both reversing lamps and the number plate. The petrol cap comes off easily and the filler neck can be masked or removed. The lamp clusters can be masked but we do know they are quite easy to remove. That just leaves the chrome bumper. What would you do?

bumper you may be more inclined to reach for the masking paper and tape.

Wrap the bumper in masking paper using a generous overlap as you spiral the paper round the bumper. When you have done as much as is practical with the paper, secure the ends with tape to prevent it unwrapping under the force of the spray gun. Next, cover exposed bumper brackets with individual pieces of tape.

How to Remove Masking Tape

Ensure that the paint is dry before attempting to remove masking tape. We hesitate to give you the next piece of advice, as you could cause more harm than good. Touch the paint lightly with a finger. If your finger comes off dry the paint is dry.

However, if the paint is NOT dry you have left a fingerprint on the painted surface! Be warned!

A safer way to decide if the paint is dry is to touch a piece of painted tape. If the paint is dry on the tape you can be fairly certain it is dry on your panel. If the paint is still wet on the tape it will be just as wet on the panel, but you haven't caused any problem.

The correct way to remove masking tape is to get one corner or edge loose, then pull the tape outwards at 90 degrees to the panel.

Having now masked the vehicle ready for painting, the next stage in the process is to mix some paint.

Mixing Sticks

Using a mixing stick is an easy form of finding the right thinning ratio. Always choose the right mixing stick for the paint your are using. The thinning ratios have already been worked out for you.

For example when thinning Cellulose or

One Ford 100E, all masked up and ready for spraying. Attention to detail and a meticulous approach to masking will be rewarded when the job is complete. Thick brown paper has been used plus wide masking ensures that nothing will come loose when the compressed air starts flow. This is the standard to aim for. Photo: Bert Wiesfeld.

Synthetic the operator will have only two columns to deal with on the stick. The first column will be the paint and the second column the thinner. If you pour paint up to the level three in the first column you will need to go to level three in the second column to achieve the correct thinning ratio.

If using Two-Pack materials you will have three columns to deal with. The first column will be paint, the second hardener and the third column the thinner.

What, No Mixing Stick?

There is a rough and ready way to thin paint without a special mixing stick. Pour a quantity of paint into the pot. Stand a clean stick or piece of welding wire in the paint. Remove the stick and the paint will remain on the stick and show the level. Now hold the stick so that the bottom of the stick is just level with the surface of the paint.

Pour in thinners until the level of the thinners reaches the paint level on the stick. This will give you a 50/50 mixture.

This can only be used as a guide and you really should purchase the correct mixing stick when buying your paint.

Viscosity Cups

The correct way to check if paint has been mixed with solvent in the correct proportions is to use a flow cup or Viscosity cup.

Viscosity cups are used for checking the thinning ratio ***after*** the paint you are using has been thinned.

The cup has dimensions to a British Standard which defines its capacity, shape and the dimensions of the hole at the bottom.

The idea is that you fill the cup with a sample of your paint and thinners mixed together, level off the sample to the top of the cup, then unblock the hole at the bottom.

This beautiful (and rare!) Vauxhall estate car would present a challenge to any spray painter. It has two-tone paintwork, (note the chrome strip covering the join), it has lots of chrome to mask or remove, plus very wide panels -- just look at that bonnet. Despite all these traps waiting to catch the novice sprayer, this car is immaculately finished, the paintwork and chrome gleam, and is a credit to its owner.

Chapter Five

Make sure you read the instructions on the tin before mixing paint. This 5 litre can carries the usual warnings plus vital information about what to do if the Cellulose thinner gets in your eyes or is swallowed. You are advised not to allow thinners to enter drains, otherwise the Fire Service must be informed.

The paint will take a certain amount of time to pour out of the cup at a given temperature. For example, at 25 degrees C paint may take 24 seconds to fully drain from the cup. If the same paint was to have further solvent added and the test run again the thinner mixture would pour out of the cup quicker.

Manufacturers specify how their paint should be thinned and in what proportions and then give information about how long the paint should take to drain from the viscosity cup.

The paint is considered to have drained when the flow of paint turns into a series of drips.

For example One-Pack Acrylics when thinned should go through the Viscosity cup in about 15 to 18 seconds. If the paint goes through the viscosity cup in less than 15 seconds then the paint is too thin and needs thickening up by adding more paint. If the paint goes through the viscosity cup in more than 18 seconds then the paint is too thick and more thinner needs to be added.

Pouring Paint

When you open a paint can and pour paint from the can into a container you will probably spill some. If your hands are clean you can probably wipe the excess paint back into the tin. Paint is expensive so don't waste any.

Another tip is to open the can, then punch a few drain holes in the trough round the can. This will allow any excess paint to find its way back into the can. When the lid is replaced, the holes will be blocked. As you become more proficient in pouring you will get lest waste.

Mixing and Filtering Paint

It is recommended that you pour paint into a separate container before pouring it into the

Paint and thinners are available in various quantities. Make sure you order enough for the job being done. For amateur sprayers that 25 litre tin may be too much. In this case a professional workshop were going to spray a commercial vehicle. Always remember to keep some thinner on one side for spray-gun cleaning.

gun. This way you can add thinners, hardener or whatever and mix it up properly before straining it into the gun's paint pot.

You can buy packets of ten paint strainers, fine, medium or coarse. If you don't have a strainer handy then a nylon stocking or tights can be used. These were used regularly in the old days -- more by people brush painting as paint strainers were not used much. These days, as strainers are so cheap, (about £1 for ten), they are used more widely.

The other way to strain paint is to attach a strainer or filter to the gun on the end of the stem. There are basically two types of these: one looks like a witches hat, which fits up inside the stem and a clip secures it in place. The other is a cup which goes over the end of the stem.

The problem with the cup on the end is that it accumulates some paint, so when you put your gun down to top up your paint pot, paint tends to run out of the strainer and make a mess. With the witches hat the paint runs straight out into the pot so you don't get the same mess.

Ensure that paint is thinned and then mixed *thoroughly.*

Basic Spraying Technique

There are very few mysteries to the actual spraying process, but even so there is plenty for the amateur to do wrong.

Firstly, hold the gun parallel to the body panel and between six to nine inches (15 - 22 cms) from the surface of the panel. For some paint systems eight to ten inches is recommended. The span of your hand is a good guide.

Get used to the gun by working the adjustment screws. If you are doing door shuts you need to reduce the fan to a minimum. If applying primer or top coats, open up the air control valve to maximum,

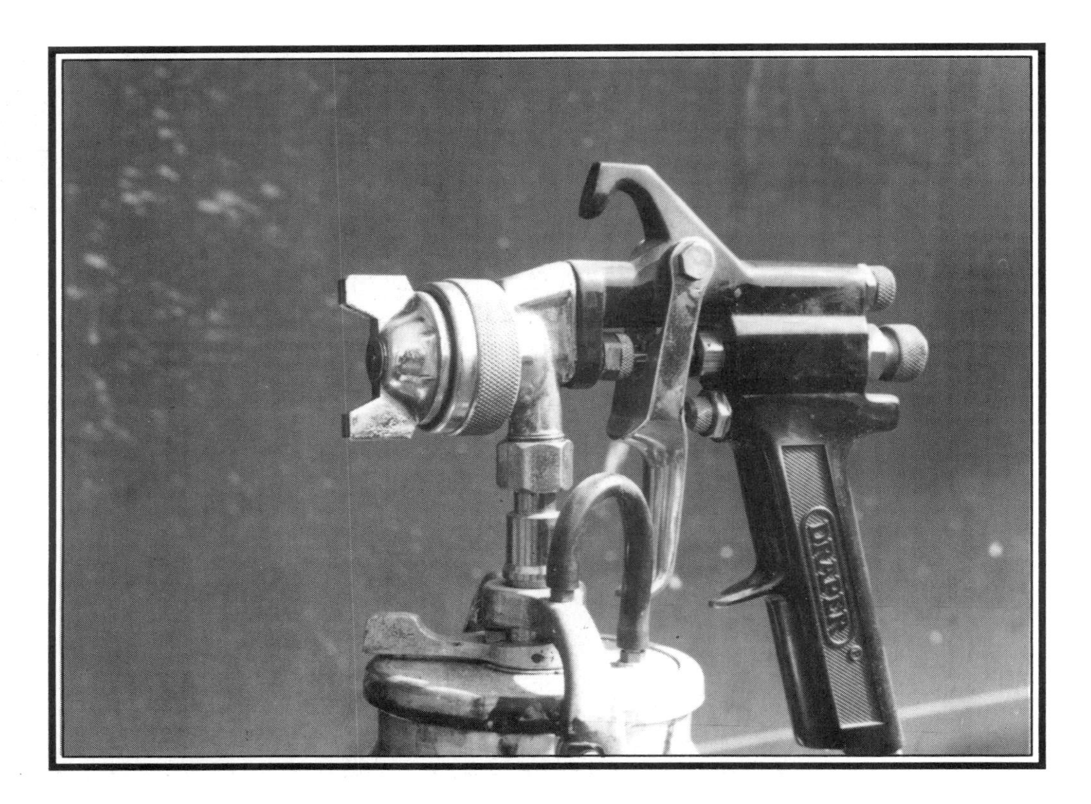

This Draper spray-gun has a plastic tube connected to the breather hole on top of the fluid container. The advantage is that when the gun is tilted, for example when spraying a roof panel, no paint will drip out. The owner of this gun keeps a few inches of thinner in the fluid container when not in use, and stores the air cap, immersed in thinners, inside the container.

giving the greatest fan spread. This should be the width of span of your hand.

The gun must be moved across the area to be sprayed and still remain parallel to the panel. DO NOT ARC. Arcing is the situation where your elbow moves the gun; it will be at the correct distance from the gun in the middle of the pass, but at either end will be well away from the panel. (The only time you don't hold the gun parallel to the panel is when doing fade-out techniques. These are covered elsewhere in the book).

At the same time, practice pulling and releasing the trigger on the gun. It is not so daft to actually hold the empty spray gun in front of the mirror!

When spraying, position yourself so that you can move your body relative to the panel. Left-handed people will find that placing themselves to the right of the centre of the panel will work best, while right-handed sprayers should be to the left of centre.

Hold the spray gun near the top of the panel and about six inches to the side of the panel. This is important! In order for good paint coverage on the panel, you must start your spraying "pass" off the panel. See the diagrams which start on page 108.

In the hand not holding the spray gun, hold the air line. You do not want it getting in the way during a pass, or worse still touching the paint you've just done.

The trigger has three positions. The idle position means the air and paint supply is shut off to the gun. Pull the trigger half way in and you get straight compressed air and no paint. Pull the trigger all the way back and you get a mixture of air and paint.

Pull the trigger on fully and start to move across the panel. I cannot tell you how fast to move. This is down to your experience, the type of paint being used and the ratio of paint to thinners. If you are too slow moving across the panel there is a greater risk of

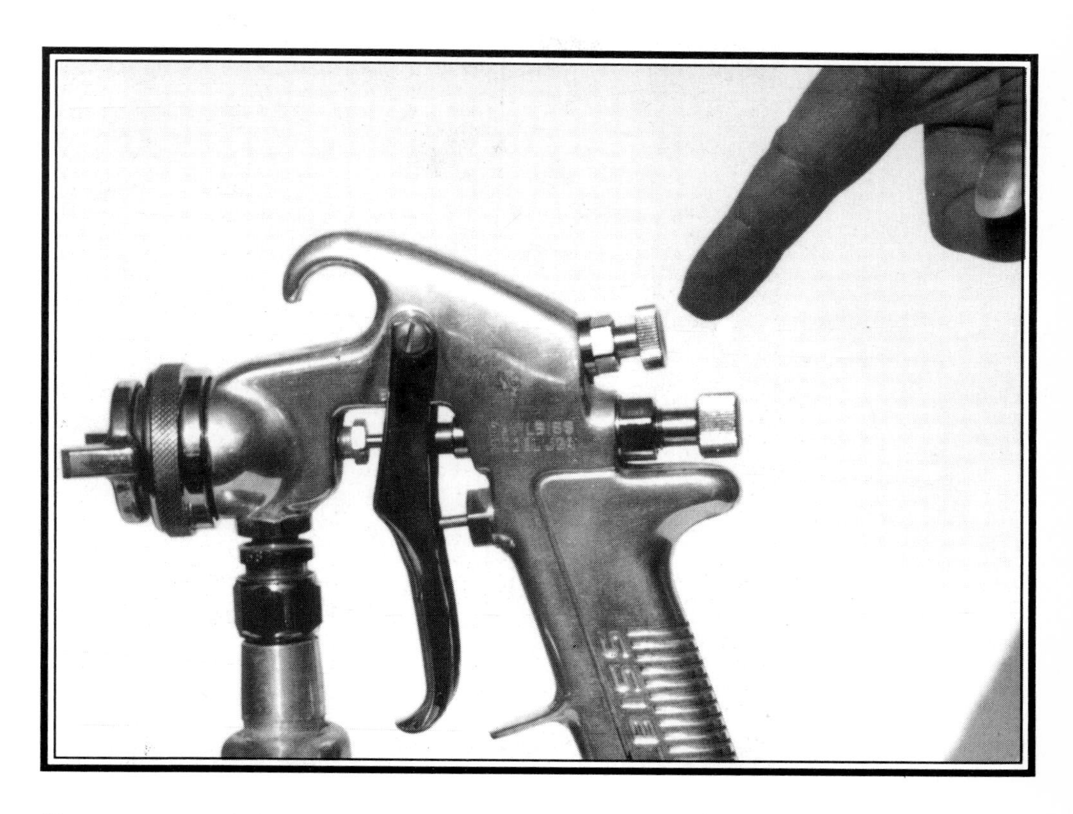

The next couple of photographs indicate the controls on a typical spray-gun. Being pointed to is the Spreader Adjustment control. This controls the flow of air to the air cap and determines the spray pattern, from a large oval spray to a small circle. See page 106.

"runs" as you are putting on more paint. Move too quickly and you will have a very thin or patchy coverage.

Finish your pass six inches off the panel. Release the trigger completely.

Move the gun down so that your next pass will overlap the first pass by about 50%. Assuming you have made your first pass from the left hand side to the right, pull the trigger fully on and start the next pass from right to left. Finish six inches over the left hand side of the panel.

Yes, I know this is wasteful of materials, but that is how it is done. As you gain more experience you may be able to reduce that six inches "off panel" and so save some material.

The aim is to get the paint on EVENLY. If you can achieve this you will not get runs. Drop 50% each pass. If you drop 100% you get wet and dry patches.

If you put paint on evenly you will not get runs. It does not matter how wet it goes on as long as you put it on evenly. Do not jerk the gun, and overlap 50%.

Continue passing back and forward and overlapping 50% each time. Be sure to cover corners and edges of panels too. Also, for the best finish, use the best thinners and the best paints. Easy to say, but it is so easy to be tempted to cut corners and save a fiver here and there. Not worth it.

Beginner's Bungles

There are several problems which the first-time sprayer will encounter when he picks up the gun. His instructor has probably said something about being confident, but that is probably forgotten in the first few minutes.

You are standing there with a full pot of paint, a hissing air supply, an unfamiliar face mask and perhaps worst of all, other people watching.

Try to remember these DO's when you pick

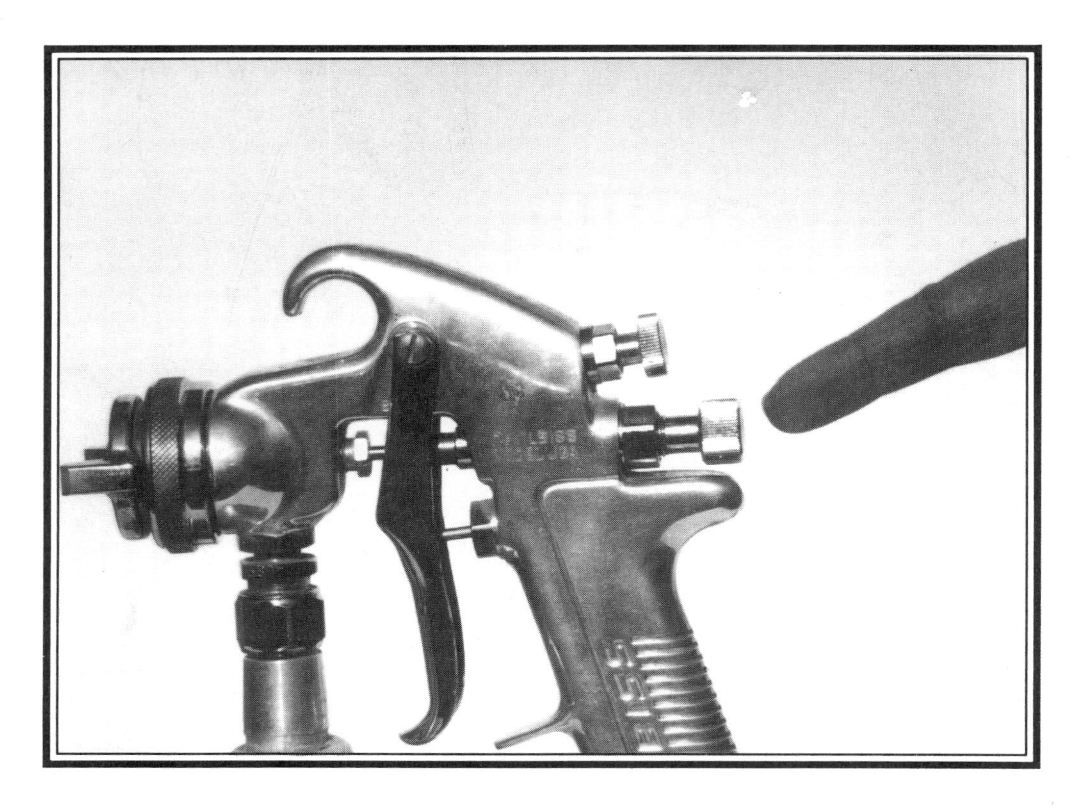

This control adjusts the fluid needle, which in turn determines how much paint passes through to the fluid tip.

up the gun for the first time to practise on an old panel:

1) Be confident. Don't be scared of the gun. If you have all the correct equipment the worst that can happen is that you "run" the paint. NOTE: If you wear spectacles you may find that they steam up when you wear a face mask.

2) Remember your triggering technique. ON before reaching the panel, OFF after you pass over the end of the panel.

3) Pull the air FULLY ON.

4) Watch for the paint going on. (The first time I ever held a spray gun I sprayed grey primer on top of grey primer. I couldn't see the paint going on.) For your first time, try to spray a contrasting colour, say black onto a white or grey panel. Better still, spray black paint onto white paper! You will then clearly see what you are achieving.

5) Remember a 50% overlap on each pass.

If you are in a class of other people and have never sprayed before, you are probably worried about making a fool of yourself. DON'T! Everyone had to start somewhere. Some people pick up the gun and spray like they did it all their lives. Others, *(like me, writes Tommy)* need some time to think it all through and get familiar with the weight, feel and balance of the gun. Everyone is different. It takes time.

Correcting Problems

Refer to page 121 if you encounter any spraying problems.

Refer to page 122 if you have any paint problems.

Dealing with Runs

If you are spraying Cellulose paint and you get a run do not worry. Wait for the paint to "flash off" (twenty minutes should be a safe time to wait), then rub the problem area with some 800 or 1200 wet and dry paper, usually used wet. This will flatten down the run which is actually a river of excess paint.

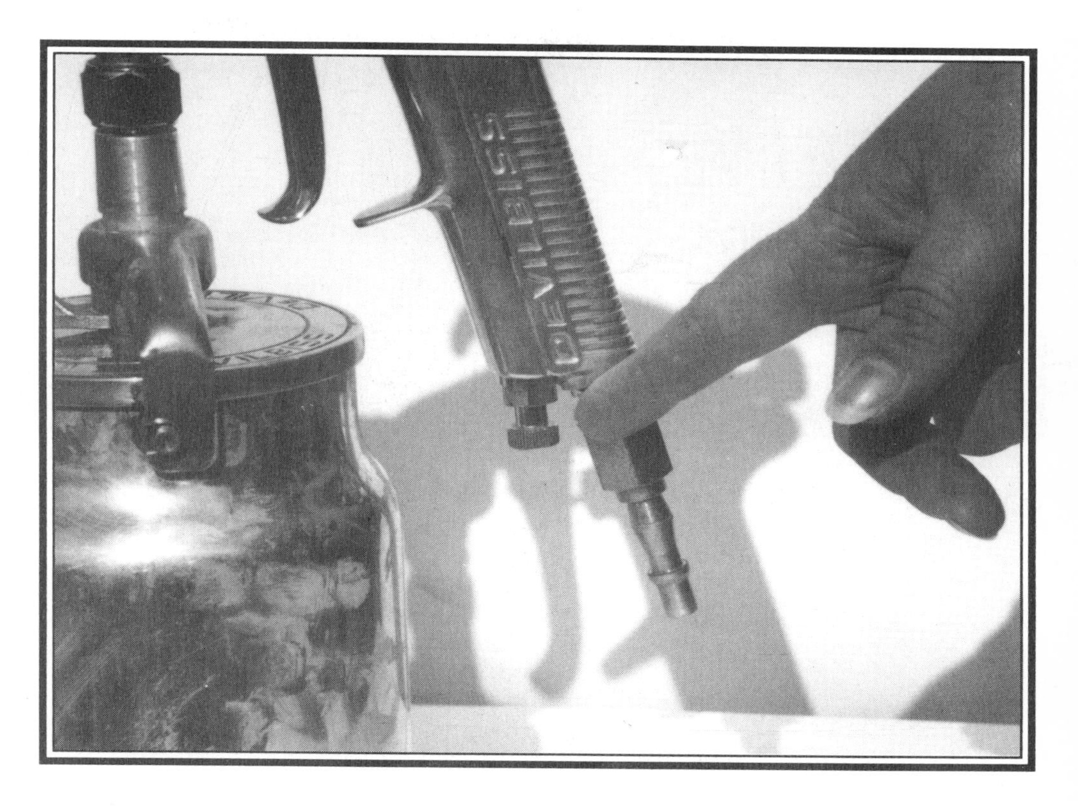

The control on this De Vilbiss may not be present on other guns. It controls the air flow to the gun, so that you can adjust the air flow at the gun quickly and easily instead of having to go to the compressor in the middle of the job. To the right of the control is a quick-release coupling for the compressed air supply.

Flatten down the run, clean up the resulting mess, then spray over the problem area.

Hopefully this time you will have corrected the error which caused the run and the second attempt will be perfect.

If spraying other types of paint, details of how to deal with runs are given elsewhere in the book.

Yes, I had to ask Trevor what the hook was for! You can hang the gun from the hook while you do something else. The gun will not sit on a bench with the air-line connected. It would just topple over!

Willow Publishing Paint Company

Willow Publishing

747

Document Control

Issue Date: 28th March 1993

Supercedes: 18th August, 1992

Specimen Data Sheet for *Good Cover* Paint

This introductory section will state what type of paint system is being discussed, plus a few words about specific features: For example: ***Good Cover*** is a Cellulose car finish which provides fast air-drying.

There may well be another heading which lists other related Processes:

Processes:

Good Cover standard process,

Good Cover high gloss process,

Good Cover low gloss process.

Intended Use

These products are for the professional painting of motor vehicles ONLY after reference to manufacturer's data sheets numbers XXX, YYY and ZZZ.

Products

This section will list all the products associated with the ***Good Cover*** paint system, such as paint, thinners, retarders, etc. together with the manufacturer's product number. For example:

WP1 ***Good Cover*** solid colour,

WP2 ***Good Cover*** standard thinner,

WP3 ***Good Cover*** rapid repair thinner,

WP4 ***Good Cover*** slow thinner,

WP5 ***Good Cover*** retarder,

WP6 Panel Wipe,

Etc...

GENERAL PROCESS INFORMATION

Under this heading will be specific information about how the product should be used. Typical headings might be:

- Substrates:

Well flatted existing paintwork if in good condition, plus various (specified) primers.

- Preparation:

Flat with P600 or finer wet and dry paper.

- Recoat

Good Cover can be recoated with itself or other Willow Publishing Paint Company products except ZZZ.

- Polishing

Polish by hand or machine using Willow Publishing Paint Company WP36 or WP79 polishing compounds.

- To Decrease gloss levels

To decrease gloss levels (perhaps for interior finishing) use WP77 Matt in the ratio 8 parts WP3 to 1 part WP77.

- Blooming

In hot humid weather use up to 15% of WP5 Retarder.

Specimen Data Sheet continued opposite...

GOOD COVER STANDARD PROCESS

Paint Type	**Solid Colour**
Mixing Ratios	**WP1 1 part** **WP2 1 part** **or** **WP3 1.25 parts** **Mix thoroughly before use.**
Spraying Viscosity	**Thinner WP2 19-21 secs BSB4 (15-17 secs DIN4)** **Thinner WP3 21-23 secs BSB4 (17-18 secs DIN4)**
Spray-gun Fluid Tip	**0.070- 0.086 in (1.8 - 2.2 mm)**
Spray-gun pressure	**45 - 55 psi (at the gun) 3.0 - 3.7 bar**
Application	**1 single & 1 double coat or** **3 single coats**
Flash-off Time	**5 - 10 minutes between coats**
Drying Time	**Air-dry at 20 degrees C** **Dust Free 5 - 10 mins** **Touch Dry 30 mins** **Into service 4 hours**
Recoat	**After "Into Service" time**
Polish	**After "Into Service" time**

At the end of the process there may be some notes on variations to the standard process. These often reflect changes required due to temperature variations:

For example:

PROCESS NOTES

Cold (below 15 degrees C) use AAA thinner

Normal (15 - 22 degrees C) use BBB thinner

Warm (22 - 27 degrees C) use CCC thinner

Hot (above 27 degrees C) use a blend of 90 parts AAA and 10 parts CCC thinners.

Chapter Five

Setting Up Your Gun

Mix some paint and thinners in the correct ratio of thinners to paint. Spray a few test patterns onto pieces of white paper.

For spraying wide surfaces, fix the air cap so that the horns are at the right and left sides of the gun. The spray fan will then be vertical.

For spraying narrow surfaces, such as "A" posts, where you do not want a wide spray pattern, either set the air cap to the "up and down" position (the spray fan will be horizontal) or use the round pattern discussed below.

Correct Patterns

The two possible correct spray patterns are shown above. The left hand one shows the spray pattern from a correctly adjusted spray gun. This is used for all general spraying. The dot on the right can be used for spraying specific areas, for example spot repairs, or narrow areas such as "A" posts.

This round pattern can be achieved by turning the spreader adjustment valve on the gun (the top adjuster on the De Vilbiss gun) until the required pattern appears.

Faults

Any other shape of spray pattern is a fault. They are not illustrated but can consist of right handed half moons, left handed half moons, shapes like the glass of an egg timer and shapes like almonds.

All of these faults can be corrected by cleaning the air holes in the horns of the air cap, cleaning the fluid tip and/ or resetting the controls of the gun. Check also that the air pressure is correct for the gun and the material being sprayed.

Refer to page 118 for details of how to clean a spray-gun correctly.

Setting Optimum Pressure

Fasten a few sheets of white paper to the wall.

Adjust the air pressure on the compressor so that you are get 30 psi (2 Bar) at the gun.

Turn the spreader and fluid needle controls on the gun so that they are fully open. If you are not clear about which way is "open" refer to the manufacturers literature.

Hold the gun between 6 and 9 inches (15 - 22 cm) from the paper. The quick and easy way to judge this distance is the span between your thumb and tip of the little finger on an outspread hand. Spray some paint.

Raise the pressure by 5 psi (0.3 Bar) and repeat the process. Spray some paint.

Repeat the process, raising the pressure by 5 psi (0.3 Bar) each time until you get the MAXIMUM spray pattern size. If you increase the pressure by 5 psi and do NOT get any increase in pattern area, drop back by 5 psi.

You now have the optimum air pressure setting.

Use a fresh piece of paper and make a very

Spray-Gun Set Ups

- **Air Cap Number 30**

Suitable for use with a wide range of viscosities and material finishes including CELLULOSE, ACRYLIC, and SYNTHETICS. Can be used with either suction feed or gravity feed guns. Suggested Fluid Tips, EX or FF. Nominal Air Consumption 12.6 cfm @ 60psi.

- **Air Cap Number 43**

Ideal for single layer METALLICS, CELLULOSE, ACRYLIC and general refinishing materials. Suggested Fluid Tips, EX, FW or FF. Nominal Air Consumption 13.8cfm @ 60psi.

- **Air Cap Number 80**

Particularly suitable for transport finishes, SURFACERS and NON-SAND PRIMERS. Suggested Fluid Tips, EX or FF. Nominal Air Consumption 14.5cfm @ 60psi.

- **Air Cap Number 86**

Specifically developed for use with BASE & CLEAR systems and medium solids TWO PACK finishes. Gives first class atomisation and fast application rates with an even material distribution throughout the pattern. Suggested Fluid Tips, FV (Gravity fed gun) FW (Suction fed gun). Nominal Air Consumption 13.7cfm @ 60psi.

- **Air Cap Number 880**

Particularly appropriate for the high speed application of heavier materials including POLYESTER FILLER, STONE CHIP RESISTANT PAINTS and TWO PACK high build primer surfacers. Suggested Fluid Tip, D. Nominal Air Consumption 14.2cfm @ 60psi.

NOTE: Fluid Tip Orifice diameters are as follows:

EX = 0.070in (1.8mm), FF = 0.055in (1.4mm), FW = 0.063in (1.6mm), FV = 0.055in (1.4mm) and D = 0.086in (2.2mm).

quick pass across the paper with the air trigger fully open. You are not looking for fine finish, rather the paint particles should make a very rough pattern on the paper.

Raise the pressure another 5 psi (0.3 Bar) and pass across the paper again.

You are looking for the finest paint particle size, in other words the paint is being atomized more. Continue this process until you have the optimum atomization.

Basic Gun Technique

The following pages take you through basic spray-gun technique.

1 BASIC SPRAY-GUN

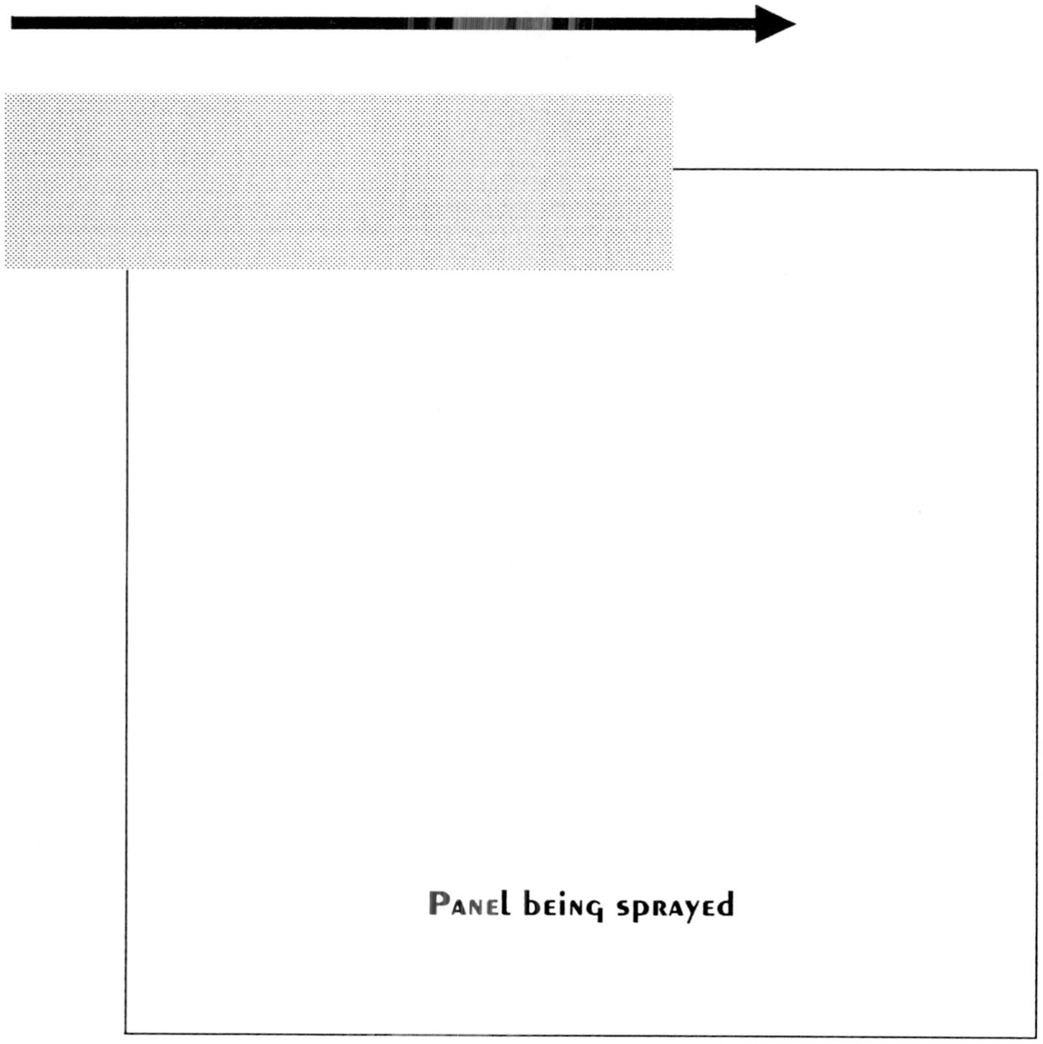

Panel being sprayed

We assume you have a panel, fully prepared and ready to be sprayed. From a starting point six inches (15 cms) off the side of the panel, pull the air trigger fully on and move the gun towards the right. The gun must be six to eight inches (15 to 20 cms) from the panel and at 90 degrees to it. The centre line of the spray fan should be approximately level with the top of the panel.

The illustration shows the first paint pass at about mid-distance.

TECHNIQUE 2

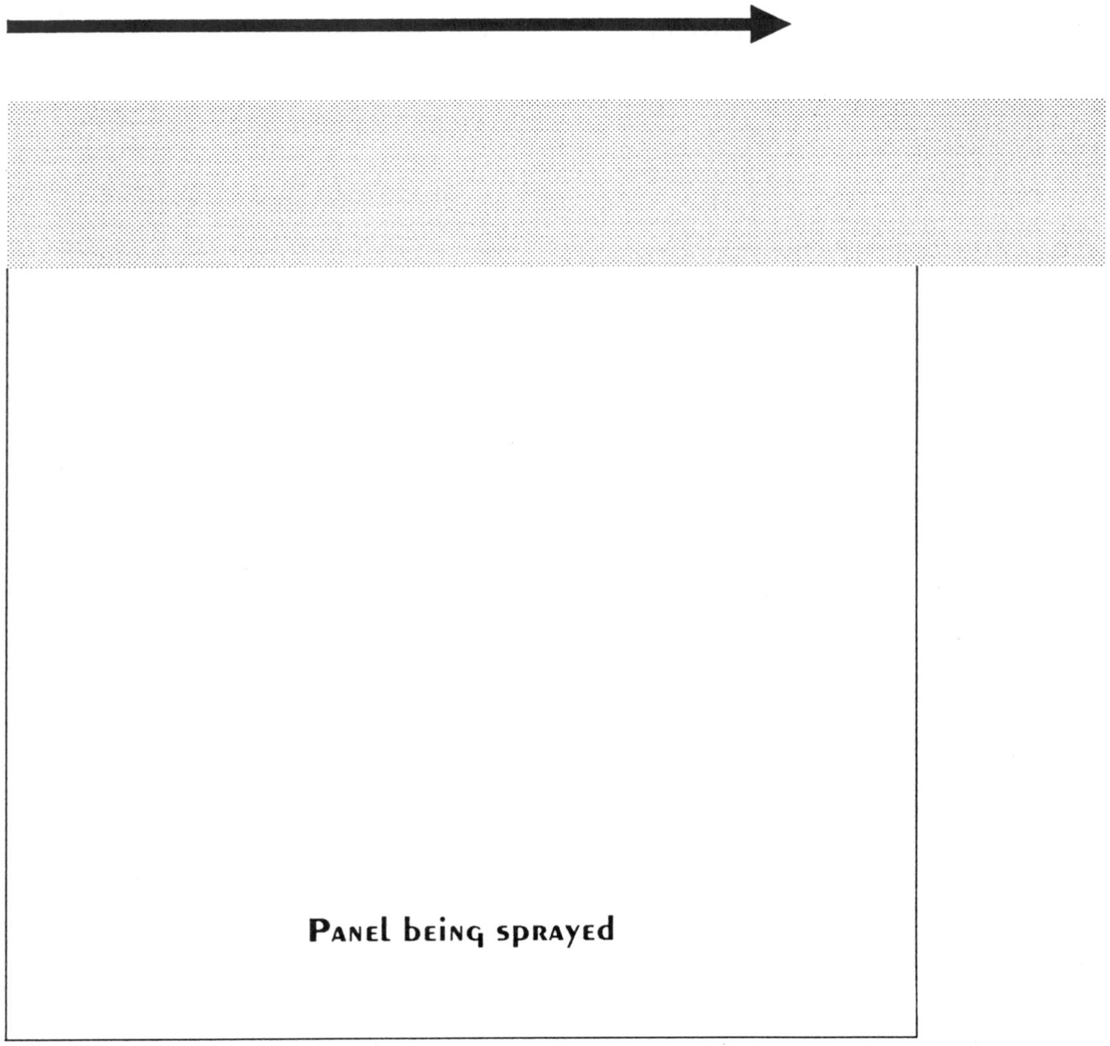

Panel being sprayed

Here the first pass has been completed. At the end of the pass the air trigger is fully released. This again should be about six inches (15 cms) off the edge of the panel.

3 BASIC SPRAY-GUN

This illustration shows the area which has now been painted.

TECHNIQUE

4

Panel being sprayed

Next, from a starting point six inches (15 cms) off the right edge of the panel, and overlapping the sprayed area by 50%, pull the air trigger on fully and begin a second pass, this time from right to left.

The illustration shows the second paint pass at about mid distance.

5 BASIC SPRAY-GUN

This illustration shows the area which has now been painted.

TECHNIQUE

6

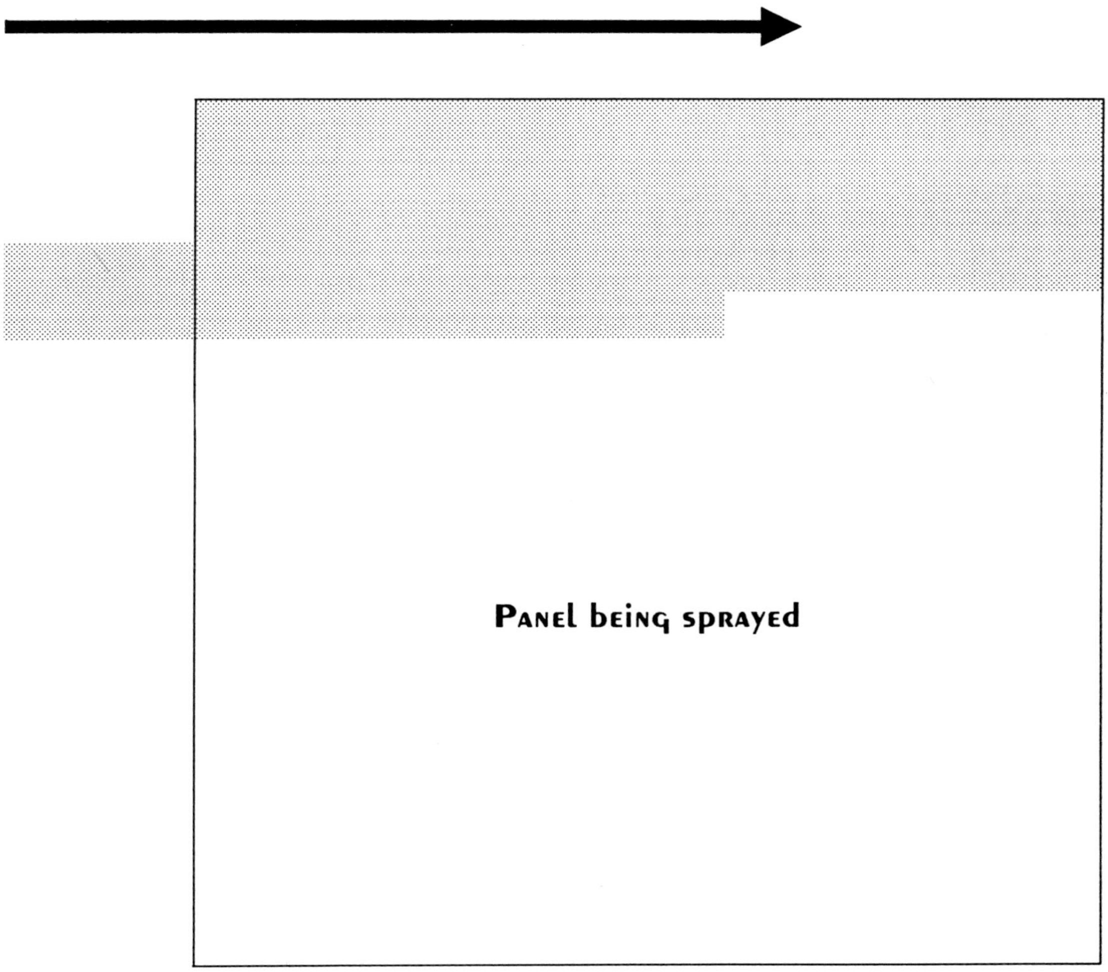

Panel being sprayed

This page shows a third pass, from left to right just beyond the mid-distance point, and again overlapping the previous pass by 50%.

7 BASIC SPRAY-GUN

After three passes with the gun, this is the area which has been painted.

TECHNIQUE

8

Panel being sprayed

Continue passing from side to side and overlapping 50% each time, until you get to the point shown above. (Remember, the final paint pass should start from off the panel, and the centre line of the spray should be about level with the lower edge of the panel. Finish six inches (15 cms) off the panel and release the air trigger. You may end up making the final pass from left to right OR right to left, but this will vary depending on the size of the panel and how well you judge your overlapping!)

This illustration shows the finished, painted panel.

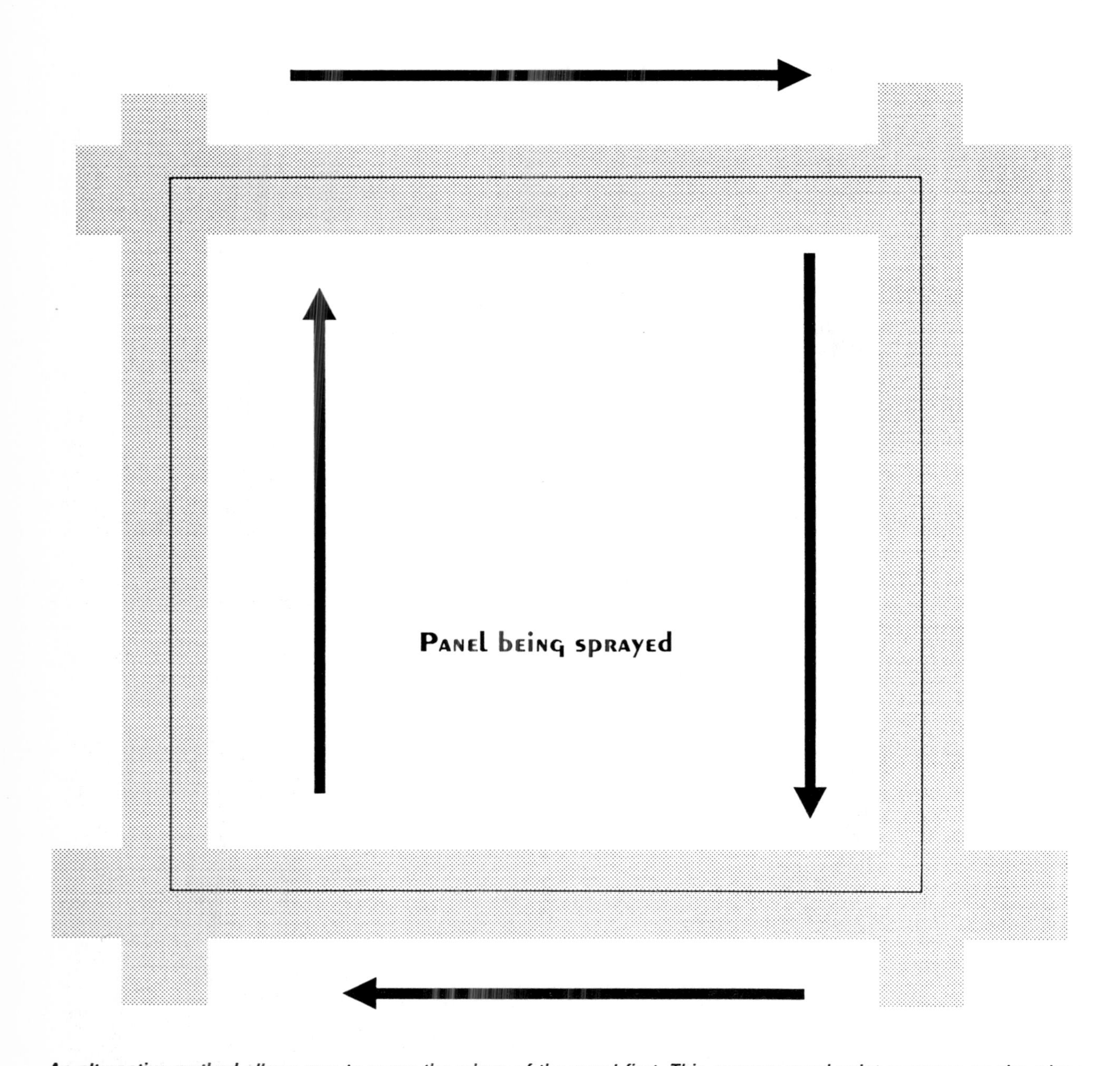

An alternative method allows you to spray the edges of the panel first. This ensures good paint coverage on the edges where paint coverage may often be thinner due to poor technique. This technique may be more useful on an irregularly shaped panel such as a wing with a lot of "shape". This page illustrates the first pass from left to right, starting six inches (15 cms) off the left edge of the panel. The centre line of the spray is about level with the top of the panel.

Next, shows the gun making a second pass DOWN the right hand side of the panel from top to bottom. Again, the pass starts six inches (15 cms) off the panel and ends six inches (15 cms) off the panel. The centre line of the spray is about level with the side edge of the panel. Continue covering of the edges, this time the bottom edge, again following the same procedure as above. In this sequence the last pass is made from bottom left to top left. This leaves the panel with all four edges sprayed. It can now be sprayed as described in the previous text.

ABOVE: This gleaming Cortina Estate was judged Supreme Champion at the MK I Cortina Owners Club national rally at Lamport Hall in July 1993. Paintwork by owner David Gaunt and son. Congratulations to both.

BELOW: Just so the big boys don't feel left out here is a beautifully turned-out Scammell Contractor dating from 1968. Paintwork needs the same preparation but more material.

Spray-Gun Cleaning

After every spraying session it is essential to clean your spray-gun. This must be treated seriously, in fact it will help you if you think of this as a surgical procedure, requiring absolute cleanliness.

Remember you can use cheap thinners for cleaning -- but NOT for spraying!

Cleaning a Suction-Fed Spray-Gun

Empty any remaining paint or solvent into a waste tin. (Remember that if you used primer straight from the tin, then you can safely pour unwanted primer back into the original tin. This may not be possible if you have thinned paint with thinners. However, consider keeping excess thinned paint in a separate clean tin for use later).

Place the spray-gun on a firm surface such as a table or bench.

Loosen the air cap by two or three turns,

Loosen the fluid cup,

Keep the fluid tube in the fluid cup and hold a rag over the air cap to block the holes,

Pull the air trigger on the gun. This will force the air through the spray gun but drive any paint back into the fluid container,

Next empty the fluid container of any paint residue and rinse out with clean thinners,

Refill the fluid container with clean thinners (about two inches of fluid will be about right).

Tighten the air cap

Spray the solvent through the gun to clean out all the fluid passageways. A tip here is to spray onto a cloth, which can then be used to wipe over the exterior surfaces of the spray-gun.

Remove the air cap and clean separately with thinners. When clean, dry the air cap with compressed air. See Cleaning the air cap, below.

Wipe down the rest of the spray-gun with a solvent soaked rag.

Cleaning a Pressure-Fed Spray-Gun

Hold some rags over the air cap to block the

Undo the lever (shown above) holding the fluid container onto the spray-gun, remove the fluid container and pour out any remaining paint or solvent. If paint is left in the container it will dry out and may clog up the fluid tube or filter.

air holes. Hold the rag firmly in place.

Pull the trigger to force air through the gun. This will force fluid back through the gun.

Clean the fluid container and fill with a few inches of solvent.

Pressurise the system and spray solvent until it sprays clean.

For further details of cleaning, refer to the manufacturers instructions.

Cleaning the Air Cap

If you manage to block one of the air cap holes with dried paint, DO NOT use a bit of wire to force the paint out of the hole. This could damage the precision-drilled hole. Instead, place the air cap in a small container of thinners and soak the offending paint out.

Unscrew the air cap and immerse it in a small cup of solvent. Dry with compressed air.

If the paint has dried, blocking the air holes, clean the air holes with a soft tool such as a match stick. Do NOT use a wire, nail or pin as this could permanently damage the precision-drilled holes.

If the surface of the air holes is damaged in any way you can try rubbing it with 1200 grade paper. This may remove any burrs or rough edges. However, the question must be asked, "How did rough edges appear on your precision-made spray-gun?"

Any serious damage can only be corrected by replacing the air cap.

And Finally

Wipe all the parts dry with a fluff-free rag. Next, oil any moving parts as recommended by the manufacturer.

Spray-guns are precision instruments and need to be treated with care. The most care you can give them is to keep them clinically clean.

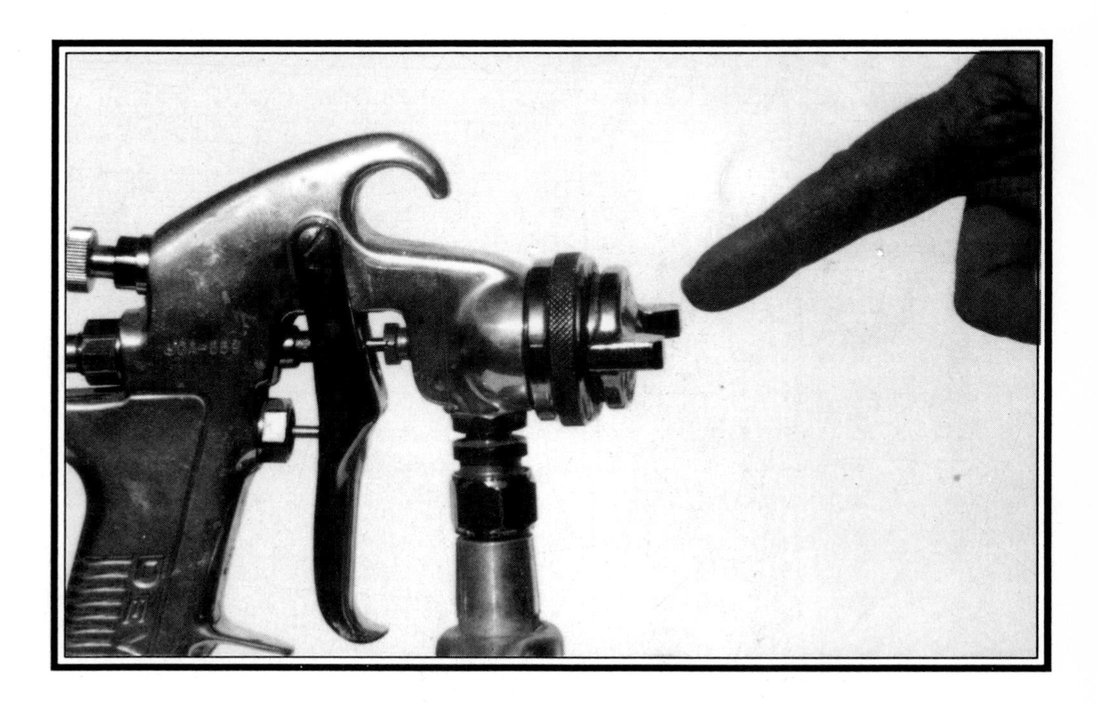

Loosen the air cap (shown above) by turning the knurled screw two or three turns. Also loosen the fluid container but do not remove it. Roll up a piece of rag and cover the holes in the air cap. Pull the air trigger fully on, and this will force any paint or thinners still in the fluid tube BACK into the fluid container. Next, rinse out the fluid container with clean thinners and pour them out. Finally, pour an inch or two of clean thinners into the container, tighten the air cap and spray the the thinners onto a rag. This will ensure the inside of the gun is thoroughly clean.

If the air cap (above) is blocked or dirty, let it soak in some thinners, then blow it dry with compressed air. Do NOT use sharp materials such as nails or bits of wire to clean a blocked air cap. This could damage the air cap.

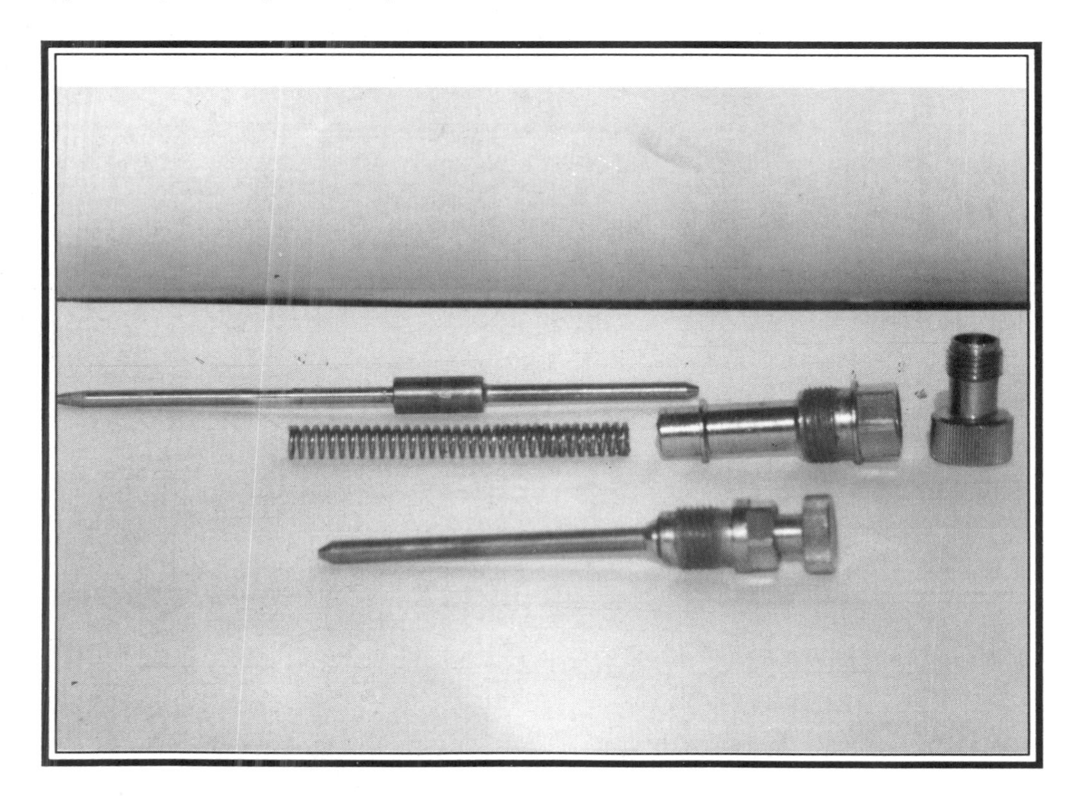

Occasionally, dismantle and clean the fluid needle and spring as shown at the top of the above photograph, and the valve assembly in the foreground of the photograph. Refer to the manufacturer's instructions for details of cleaning and lubricating.

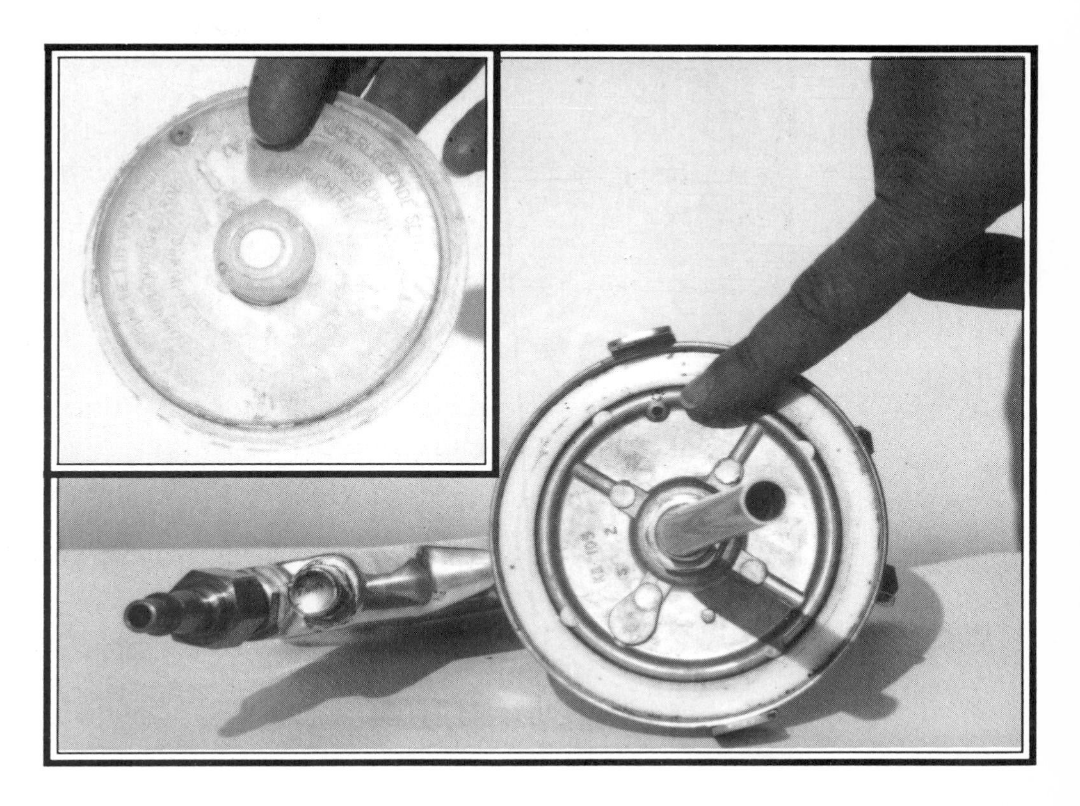

This hole (shown above) must be kept clear otherwise the spray-gun will not work properly. If the gun is tilted when the fluid container is full, paint could leak from this vent hole. A plastic anti-drip washer (INSET) will prevent this occurring. The vent hole on the washer must be located at 180 degrees from the vent hole in the gun's body.

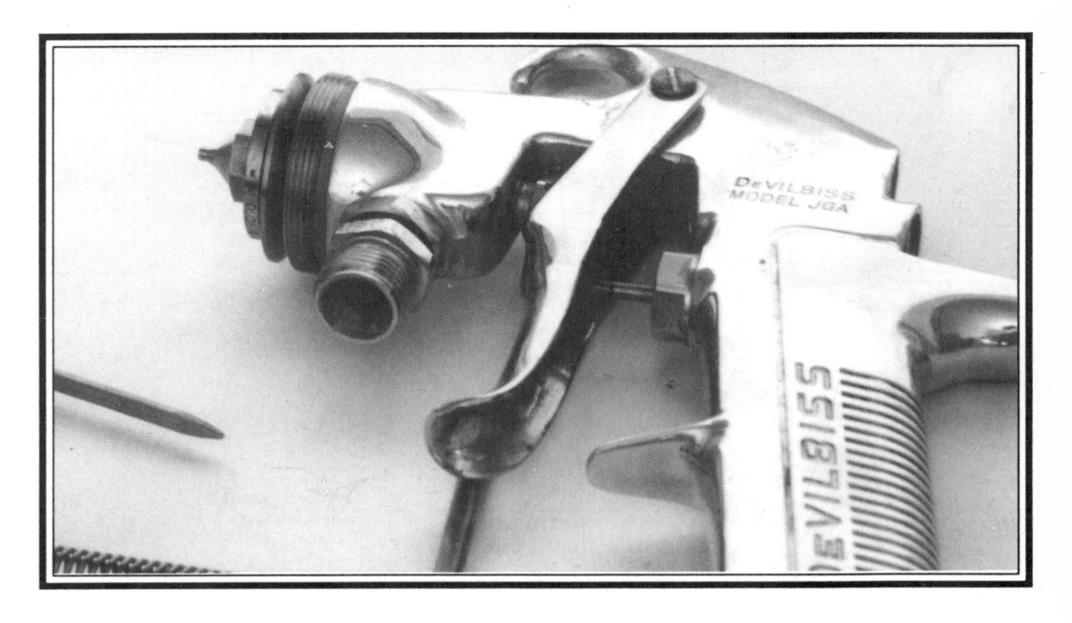

If further dismantling is needed (as above) then contact your local dealer for details of kits required during an overhaul. Your spray-gun is a precision instrument which must be kept clean and well adjusted if you want the best performance from it.

Basic Spraying Faults

PROBLEM	CAUSE	CORRECTIVE ACTION
Dry spray	a) Air pressure too high,	1) Reduce air pressure,
	b) Paint not thinned enough,	2) Mix correctly,
	c) Gun too far away from panel,	3) Spray at 6 - 9 inches,
	d) Gun out of adjustment,	4) Adjust gun.
Uneven pattern	a) Too little paint through gun,	1) Adjust fluid flow,
	b) Low air pressure,	2) Increase air pressure,
	c) Gun moved too quickly.	3) Improve technique!
Incorrect pattern	a) Fan adjust stem not seating.	1) Clean or replace.
No spray from gun	a) No air pressure,	1) Check air supply,
	b) Incorrect air cap fitted,	2) Change to correct air cap,
	c) Fluid tip not open sufficiently,	3) Open/adjust fluid screw,
	d) Fluid too heavy (suction guns),	4) Thin fluid if appropriate.
Fluid leaking from gun	a) Packing nut loose,	1) Tighten/replace packing.
	b) Packing worn or dry,	2) Replace or oil packing.
Dripping from tip	a) Dry packing,	1) Oil the packing,
	b) Sluggish needle,	2) Oil the packing,
	c) Tight packing nut,	3) Adjust packing nut.
Excessive overspray	a) Air pressure too high,	1) Reduce air pressure,
	b) Gun too far from panel,	2) Spray at correct distance,
	c) Improper spraying technique.	3) Use correct technique.
Excessive fog	a) Too much thinner,	1) Remix paint & thinners,
	b) Thinner drying too quickly,	2) Remix paint & thinners,
	c) Air pressure too high,	3) Reduce air pressure.
Thin, coarse finish	a) Gun too far from panel,	1) Spray at correct distance,
	b) Air pressure too high,	2) Reduce air pressure,
	c) Thinner drying too quickly.	3) Remix paint & thinners.

Paint Problems

Almost all painting problems are self-inflicted.

Remember these six rules and you will not encounter any of these problems.

1) Preparation and cleanliness are essential. Spend 80% of the time on preparation and 20% on actually spraying.

2) Use the best possible equipment, and ensure that it is set up correctly for the paint system being used. Pay particular attention to gun settings.

3) Use one manufacturer's paint for each job. Do NOT mix paints from different manufacturers, or change to a different maker's thinners half way through a job. Stir paint thoroughly and mix the recommended ratios of thinners to paint.

4) Ensure existing paint will work with the new paint. If in doubt test a small area first.

5) Ensure spraying area is CLEAN, has adequate ventilation, temperature and lighting.

6) Finally, ensure you have the optimum air pressure. Too high an air pressure will cause Cobwebbing, Dry-spray, Orange peel or Low Gloss. Too low an air pressure will cause Floating, Popping, Pinholes, Runs and Sags.

The following is an alphabetical list of painting faults and suggestions on how to cure the problems.

Adhesion Loss

Also known as Flaking or De-laminating

The newly applied paint detaches itself from the primer, or the metal surface. Usually this is caused by poor preparation in that the surface has not been properly cleaned and/or de-greased. This also applies if a layer of moisture existed on the panel prior to spraying.

It could also be caused by incorrectly mixed paint or the wrong thinners.

Check all these points, flat back to bare metal or primer and start again.

This problem is particularly noticeable when spraying aluminium. Here a special etching primer must be used to key the paint to the aluminium surface.

Bleeding

Also known as Staining or Blushing.

Put simply, pigment from the original paint shows through the new paint. There is a more technical explanation for this condition, which usually occurs on red or maroon colours. Seek advice about this problem. If you are in any doubt about bleeding, test spray a small area before doing the whole vehicle.

The ultimate cure for Bleeding is to sand back to bare metal, removing both the original and new paint. Then, if required use a sealer or bleed inhibitor before repainting.

If bleeding occurs, try this: apply a coat of black over the area causing problems and then repaint with the original colour. This may work and is an easier method than stripping back to bare metal.

Alternatively, if you have only sprayed one colour coat you may be able to seal the bleeding with a bleed inhibitor.

Blistering

Also known as Micro-Blistering.

Bubbles, or blisters appear in the paintwork. The cause is moisture under the paint. Poor preparation is again the cause of blistering. Oil, grease, or sweat from fingers can contribute to blistering.

Other causes include incorrect thinners or not allowing sufficient time between applying coats. Allow paint to flash off according to manufacturer's instructions.

Its back to bare metal again, either a localised area or the complete panel, and ensure thorough preparation and cleanliness. Apply paint again.

Blooming

Also known as Bleaching.

The surface of the paint appears white or milky, or in very bad cases appears to have a chalky deposit.

Caused by damp, wet conditions and is the most common paint problem for the amateur.

Can also be caused by too high an air pressure, causing cooling or thinners which flash off too quickly.

Spray in warm weather or arrange a safe form of heating for your workshop -- NOT an open flame source!

Correct the problem by rubbing down with very fine paper and respray the complete panel again.

Cracking

Also known as Checking, Crazing, or Splitting.

Shows up as a large number of tiny cracks in the paint surface. Can be shallow or deep. Old paint can crack due to age, and exposure to the elements over long periods.

Cracking is not so common in new paint. If you intend respraying an old vehicle which shows signs of the paint cracking, you may have to consider removing all the old paint and working back to bare metal.

Can be caused by too heavy an application of paint, or insufficient time allowed for the coat to flash off. Ensure paint has been mixed properly.

Sand the problem area, probably down to bare metal and start again.

Cratering

Also known as Fish Eyes, Cissing, or Saucering.

Painted surface shows small craters. Caused by a contaminant such as wax or silicone polish, oil, or grease.

Panel to be sprayed should be absolutely clean. Wipe down with a recognised degreaser prior to spraying. Do NOT touch a clean panel with fingers, as the sweat on your fingers could cause cratering.

Correct by rubbing away all the affected area, clean all surfaces thoroughly and try again.

Dirt

Caused by dust or other contaminants being trapped under the paint. Could be caused by contaminants in the paint.

Caused by poor preparation, carelessness when preparing the paint, or insufficient cleaning of the spraying area.

Dirt in paintwork can, if there is enough paint covering the surface, be flatted with superfine wet and dry paper and polished back to a shine. If there is not enough paint, flat back and spray again.

Dry Spray

Noticed as a rough surface on the finished paintwork. The usual reason for dry spray is holding the gun too far away from the panel. The solvents are already drying out when they reach the panel, so the paint does not "flow out" properly.

Could also be caused by too great an air pressure, or too thin a mixture of paint/ thinner.

Can also be caused by too high a spraying temperature, where the paint actually starts to dry as it hits the panel.

Sand down with a fine grade of paper and spray again.

Etching

Caused by a contaminant such as bird droppings, detergents, even pollution from a factory chimney.

Either flat off the problem area and polish, or if the problem is severe, flat off and respray. (You may find that if the paint or panel is contaminated that it will not polish out satisfactorily. The problem will disappear when polished, only to return several hours or even days later).

Flaking

Flaking is usually the result of an earlier defect such as stone chipping, blistering, cracking. Atmospheric conditions attack the weakened areas and cause poor adhesion.

Strip back to bare metal and start again!

Loss of Gloss

Caused by surface imperfections. Can be caused by many things including exposure to the weather, or sinkage. See, Sinkage.

The cure is to polish. In extreme cases you may have to flat and respray.

Mapping

Also known as Ringing.

The shrivelling of the edge of a repaired area, so that an outline of the repair, which has sunk, shows through the top coat.

Caused by solvents attacking the edge of the repair.

Flat back to the repaired area and prime carefully.

Mottling

This condition shows up as spotty or dark areas on metallic painted surfaces. Generally caused by incorrect gun set up, for example too high an air pressure, the spray pattern is too narrow, or the gun is being held too close to the panel.

The only cure is to re-flat the surface with an appropriate grade of wet and dry paper, and apply a new top coat.

Orange Peel

Shows as an uneven surface on the newly painted surface, usually described as looking like the surface of an orange. The cause is the paint not "flowing out" properly. Check adjustment of the gun, air pressure and your technique.

Could also be caused by too high an air temperature in the painting area.

Sand back with 1200 grade paper and polish, or if the orange peel is bad, sand and repaint the surface after correcting any gun or technique problems.

Overspray

Shows up as dry or nearly dry paint from the spray gun on the painted surface. Also appears on surfaces not being painted and not masked properly. This can be prevented by masking the parts of the vehicle which are not to be painted.

Can be caused by too high an air pressure, too thin a mixture of paint/thinners or incorrect technique.

A condition known as rebound can occur where paint mixture bounces off the panel being sprayed and drifts around, landing on some unwanted position.

Overspray can usually be polished away.

Pickling

Also known as Wrinkling, Puckering or Shrivelling.

Condition where the surface of the paint wrinkles like the surface of a prune.

Usually caused by a reaction between two different paint systems.

If you did not test a small area for compatibility, then you will have to remove the newly applied paint and either sand back to bare metal or apply an isolator, if one is available.

Poor Opacity

Where the original paint shows through the new paint. See also, Bleeding, in this section.

Generally caused by too thin an application of paint resulting in insufficient coverage.

Flat off and repaint.

Runs

Also known as Sags, Dribbles, or Curtains.

If you don't get some runs when learning to spray, you are either a gifted, natural sprayer (unlikely!), or you are very lucky.

Runs are caused by too much paint being applied to a given area in a given time. For example, if you move the gun too slowly across the panel, you will apply too much paint. Also, if the paint mix is too thin, then runs will be encouraged. Allow, plenty of time between coats of paint, mix the paint correctly, get the air pressure right, hold the gun at the correct distance from the work and finally, use the correct thinners for the job.

Allow the paint to dry, flat off. It may not be necessary to apply another coat, but you may want the practice!

Scratches

The paint shows sanding marks. Caused by

incorrect preparation of surface. Perhaps too coarse a grit was used and not smoothed down with finer grit papers. Also caused by too thin a paint to thinner ratio, giving incorrect coverage of the surface.

The cure is to flat back to the rough surface and prepare it again, using finer paper. If necessary a heavier coat of primer may be required.

Sinkage

Also known as Shrinkage.

Where the finish loses gloss as it dries and shows contours of the surface underneath. Can occur up to several months after respray.

Is paint being sprayed onto a vehicle which has already been spray several times, that is, it has a high build up of paint already.

Rub down with 1200 grade paper and repaint. If a high build up exists, sand back to bare metal.

Slow Drying

The paint takes a long time to dry. Caused by incorrect conditions in the spraying area, such as too low a temperature, or insufficient ventilation. Use of a slow evaporating thinner can add to this problem.

Could also be caused by too heavy a coat of paint followed by application of second layer before the first has dried properly.

Check your technique, spraying area and temperature. If the finished job is not up to required standard, sand with 1200 grade and repaint.

Soft Paint

Can be caused by insufficient catalyst in two-pack paints. Correct the mixing ratios, sand off paint surface and spray again.

If using synthetic paints it could be caused by a lack of driers in the paint.

Solvent Popping

Also known as Pin Holing.

Looks like little lumps in the paint with tiny holes at the top of the lump. Usually caused by too heavy an application of paint in one coat. (Check gun set up, and mix ratio of paint to thinners). Can also be caused by applying too much heat to panel while the paint is still drying.

The holes are caused by the solvent "popping" through the drying paint film.

The cure is to allow plenty of time for this excessive paint film to dry out, rub down the surface again and re-apply paint.

Water Spotting

Also known as Water Marking.

Occurs where water evaporates from a painted surface. An outline, which may be white in appearance, is left behind. The most usual cause of this is where the amateur sprays the vehicle outdoors and gets it wet by rain. Can also be caused by excessive wax on the metal surface.

Always make sure the painted area is fully dry before taking the vehicle out of the spray booth into the open.

Wet sand the problem area and repaint.

Wrinkling

Also known as Puckering or Shrivelling.

This problem is caused mainly be applying too many coats and poor drying conditions.

Appears as a shrivelling or wrinkling on the surface of the paint. It is caused by problems in the drying process.

Does not occur so often with synthetics as the drying process is so slow. If it does occur in synthetics it may be because synthetics dry from the surface downwards, the surface layer may be dry while lower layers are still wet. This distorts the surface layer.

Check spray area temperature, seal draughts, check correct mixture of paint to thinners.

Allow paint to dry thoroughly. Sand off affected area and repaint. If the problem is severe, rub back to bare metal and start again.

Wax Retention

See Cratering, in this section.

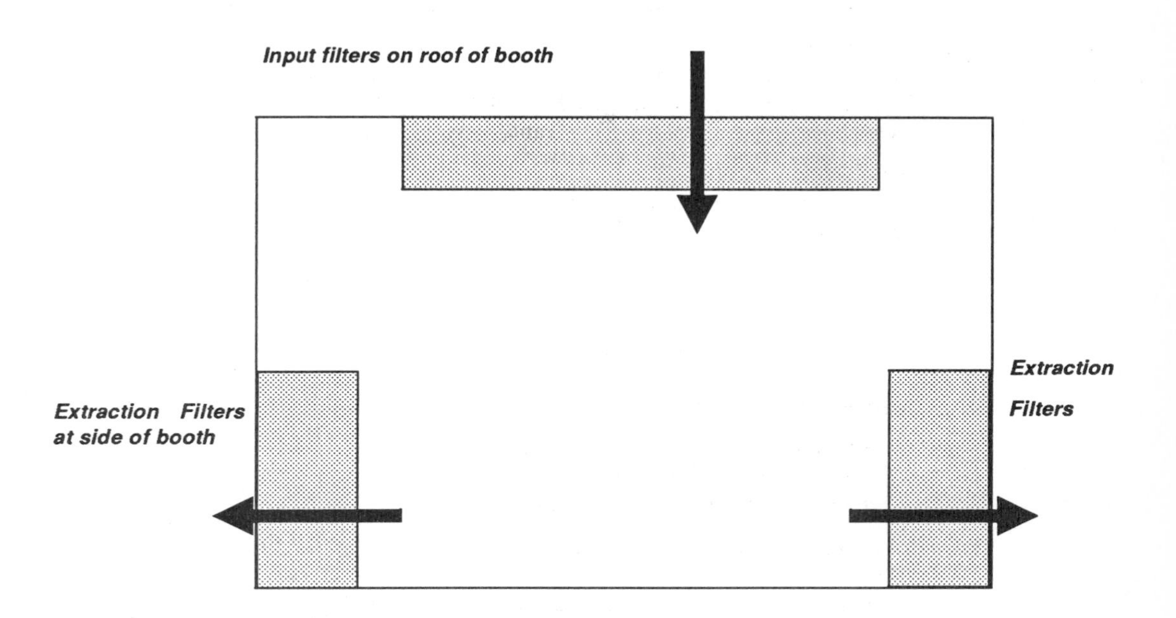

The dry extract type of spray booth, above. Air is drawn in at the top through a filter, and is extracted at the side of the booth.

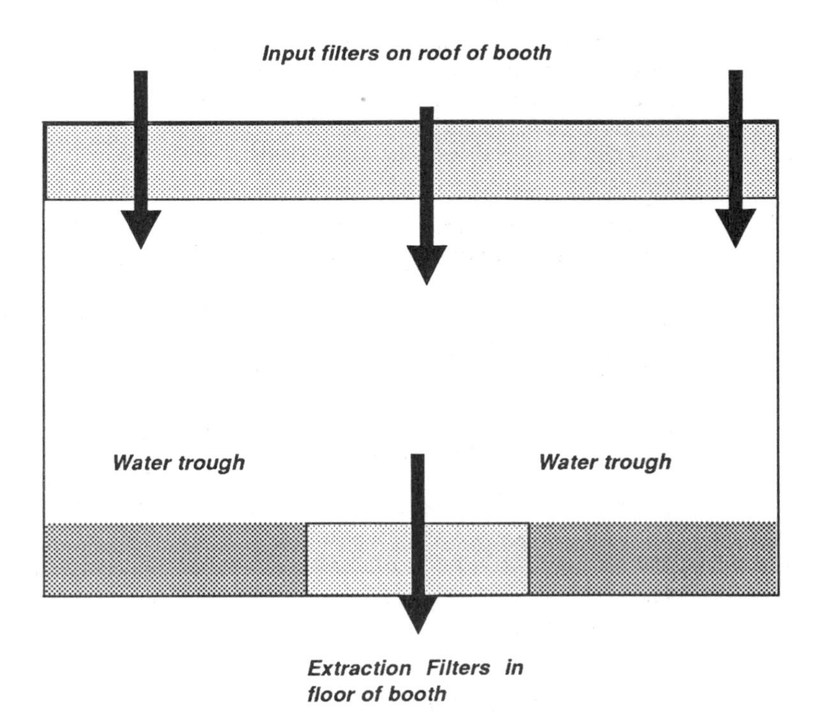

The purpose of both the dry extract booth, and the water wash booth, is to extract the spray dust as quickly as possible thus providing a dust free area. Amateur sprayers may not have the luxury of a professional booth with extraction and heating, but you can still do a professional job if you use a home-built booth and a lot of attention to detail.

Mixing, Matching & Miscellaneous

This Chapter deals with paint mixing, paint matching and other miscellaneous subjects aimed at making your respray more successful.

Colour Matching

This is a most important subject when spraying a vehicle. Get it wrong and you could be in a hell of a mess!

The first thing you need to do is look at the vehicle and make sure it has not already had a respray. For example, if you have a Ford then look under the bonnet and locate the paint code on the vehicle's Identification plate. This is the first step in identifying the paint. Take a note of the code, which will be a letter possibly followed by a number depending on how old the car is. For example a paint code on a modern car may be H 0, where the 0 means 1970 or 1980 or 1990 depending on the registration year of the vehicle. The H indicates the colour code you need, for example "Solar Gold".

Now, if the car is painted in a blue metallic, you will realise straight away that it has been resprayed from the original Solar Gold colour. You then have the task of finding out from the previous owners who did the respray and the identification of the colour used. Failing that, the paint supplier may be able to help you match the colour from his extensive array of colour chips. (A colour chip is small piece of plastic or paper which identifies a particular colour. There are probably thousands of colours available, some will be easy to match, others very difficult).

If the colour on your car appears to match the colour code, then go to the motor factor or paint supplier and get them to confirm your colour against the paint code. One good way to do this is take a little piece of the car with you! It only needs to be a small item, such as a petrol filler cap (more applicable to an old car) or a chip of paint removed from the car.

Get this cross-matched with the paint supplier's colour chips and when you are satisfied about the colour match, ask them to mix the required quantity of paint to that specification. Unless you have a really good eye for colour, then you will have to rely on their experience to mix it to the original colour. (Remember it is better to have too much paint than too little. If you have too little and have to order more you run a slight risk of the second batch being slightly different from the first batch).

You need to find out what paint system has been used, for example Synthetic, Cellulose, or Two-Pack. Refer to Chapter One for details of these paint systems. Beware, if you think you might have Synthetic!

Colour Variants

Like most things in life an easy job often becomes difficult. You think you have successfully matched the paint and are ready to order. But wait!

Paint manufacturers often tend to put out colour ***variants***. If you think you have a colour variant, you MUST find out which variant you have.

You may have a red car for instance. There may be four or five shades available. If you are only spraying a few panels, then your shade, freshly mixed from the paint supplier, may be different from the original and they won't match, with disastrous results.

If you are choosing a new colour for your car and you encounter the "variant" problem, choose the paint with variant Number One. This will result in fewer problems if you ever

If you are spraying a stripe on a door panel, as indicated above, then spray the stripe FIRST if possible. Then, when you come to spray the other colour you only have to mask the smaller area of the stripe. If applying a coach line then read the text to make sure you get it straight! There is nothing worse than a zig-zagging coach line.

Two stripes down the bonnet is a very popular enhancement to a car. They can be parallel, as shown above, or can converge at the front. Be careful if you spray a white stripe OVER red for example. The white stripe will become pink! Always spray the stripe (or stripes) first as discussed in detail in the main text.

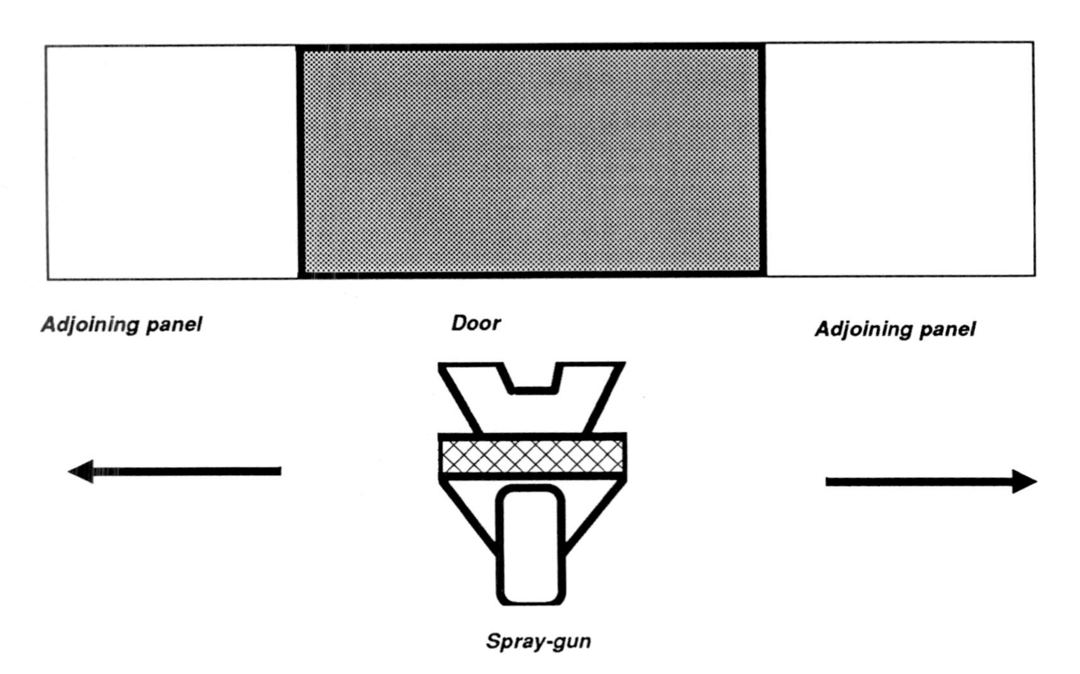

Fade-Out Techniques. The first coat must cover the repaired area. Keep the spray-gun parallel to the panel.

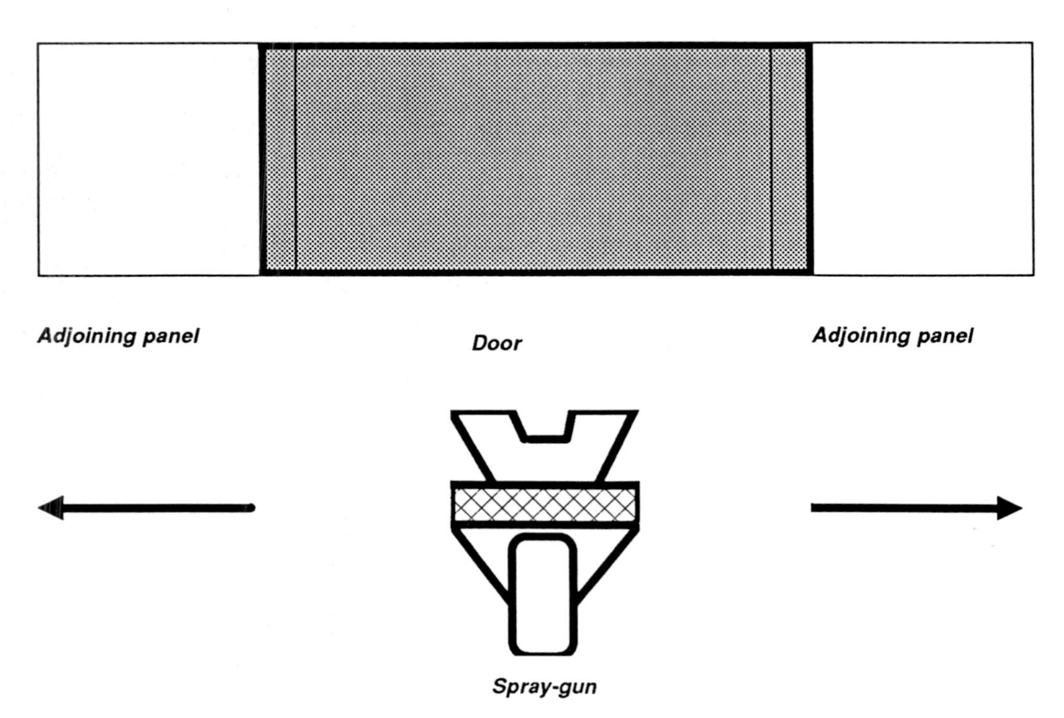

The second coat covers the repaired area plus a little bit further out over the edges of surrounding bodywork. Keep the gun parallel to the panel.

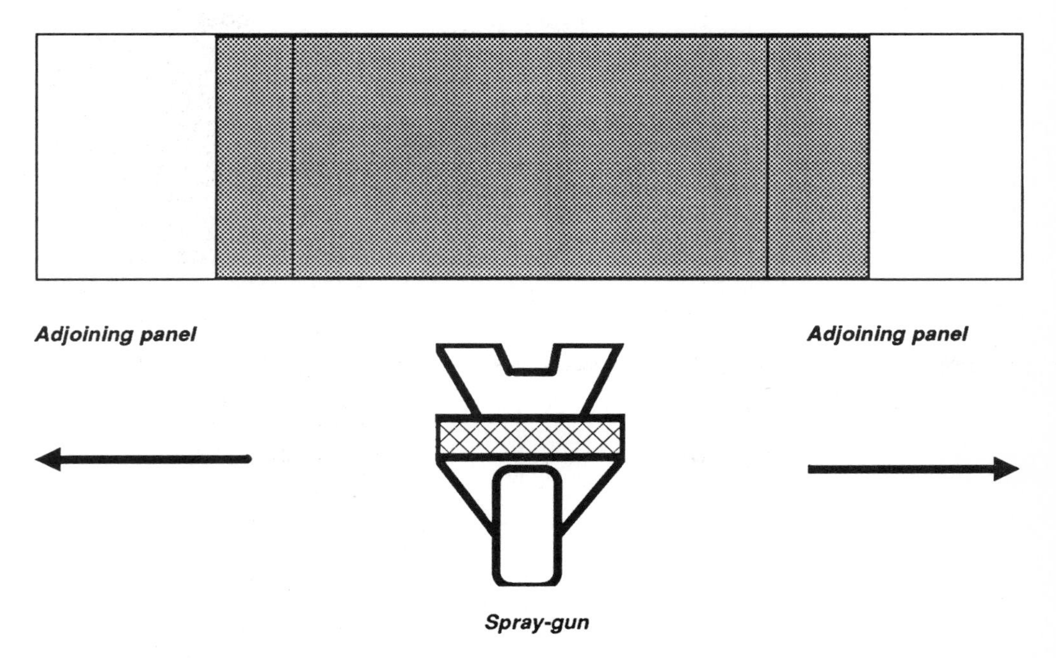

ABOVE: The third coat (which could be the final coat if all has gone well) is spraying the repaired area plus a little bit more of the surrounding area. Yet again, keep the gun parallel to the panel.

BELOW: Finally, thin the paint, lower the air pressure a little and arc the gun as illustrated to achieve a good fade out. Arc the gun from the centre of the pane so that the edges receive a very light coating. This is a skill which does need practice.

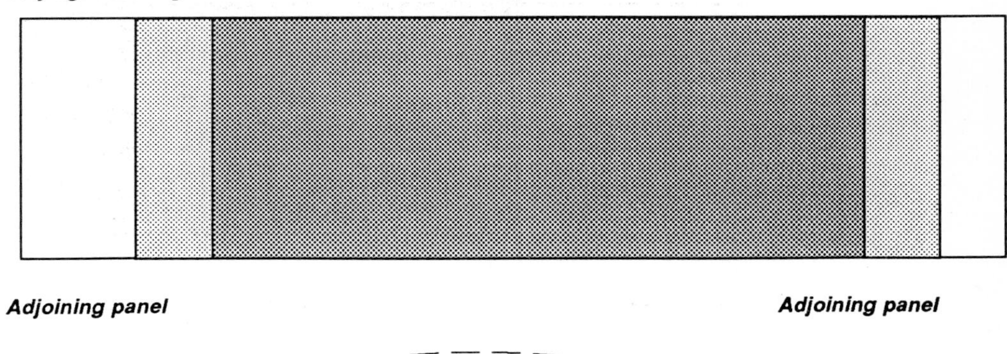

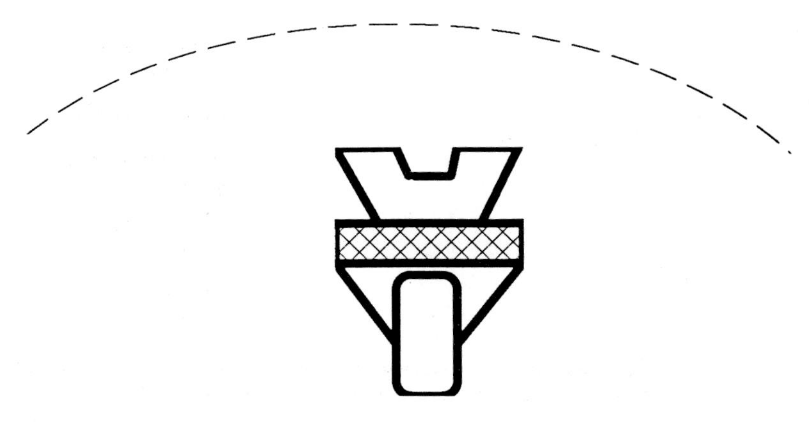

On two tone vehicles like this Ford Prefect you may follow expert advice and spray the top of the vehicle first. Spray OVER the line where the colours will meet. Next, carefully mask off the top of the car and spray on the darker colour. To save blending-in problems it helps to have a chrome strip or other decoration to soften the join.

have to match that colour again, for example if you are repairing some accident damage.

Once you have overcome this problem, the matching process should go smoothly. If you do have a slight colour mismatch then you can do some manipulation. Refer to the Section on Metallic Manipulation in Chapter Ten.

Which colour to spray first?

If spraying more than one colour on a vehicle, for example a white roof on a green car, you may not know which colour to spray first.

One general rule is to always spray the top of the car first! That way, if any further rubbing down has to be done, the dirty water won't be running down over new paintwork.

Some authorities state that the light colour should always be on the top of the vehicle, unless you want to "flatten" the appearance of the vehicle. If the dark colour is put on top it tends to make the car look shorter (that is, less tall) than it actually is.

One thing to be careful of when spraying several colours is not to build up doorsteps on the middle of the panel. For example if you spray a few coats of the base colour, then mask off certain areas for another colour. If you spray too many coats of the second colour you will create a step between the original colour and the second colour.

This may not be a problem, but if the situation is exaggerated you can get quite a large step. This situation may occur when spraying a white panel on the door of a competition car for example.

Adding a Stripe

If you are spraying a stripe on a vehicle, for example, a "Starsky & Hutch" design where you have a red vehicle and a white stripe, then it is essential that you spray the white stripe FIRST.

If you sprayed the car in red then sprayed the white stripe you would have the red bleeding through the white. The white would come out a pinkish colour.

If you want to add a stripe to an already painted vehicle, perhaps to flash the car up a bit, then you need to proceed as follows:

Mask off the areas surrounding the proposed stripe. For example, two parallel strips of masking paper will give you a stripe of the new colour. The stripe could be a single one along the side of the car, or two stripes down the front of the bonnet, or whatever you choose.

Mask off the remainder of the vehicle to prevent overspray damaging existing paintwork.

Next "Scotchbrite" the areas between the masking tape, that is, the area where the stripe will be.

Then put colour on, whether it be Two-Pack, Synthetic, Cellulose or whatever.

Ensure that the vehicle is NOT painted in Synthetic before you apply Cellulose or Two-Pack, otherwise you may encounter "Pickling" problems. (See the Glossary which starts on Page 9).

Full Respray with Stripe

If you are doing a full respray and intend to feature a stripe or stripes, then it is always easier to do the smaller areas first. So mask off the stripe area, spray on the stripe, then when you come to spray the rest of the car you only have to make off the relatively small area of the newly painted stripe.

If you choose to do things the other way round, you will have to mask off most of the vehicle before you can spray the stripe, as described in the previous paragraphs.

Where a stripe, as on both these MKI Cortinas, is to be sprayed, make sure that you spray the stripe first. That way you will not get any "bleed" problems where the main colour shows through the stripe. It is also much easier to mask off the smaller area of the stripe before applying the main colour coats.

Chapter Six

Painting Hard Plastics

Here we are talking about spraying onto a hard plastic such as a Maestro front panel, Ford front bumpers or skirting panels.

If the plastic is not in colour already, that is just in a black or grey coating, we need to start with a few coats of a universal primer which is a clear primer.

Spray one or two coats on, and leave about an hour to flash off. Do NOT flat, just leave it to flash off.

Then we are ready to spray on the colour, which needs to be a Two-Pack material, so you get a bit of movement in the material. There is no point in putting in any additives other than the hardener and the thinner for the paint. There is no flexibility.

Do NOT use Cellulose. It may look all right but it dries very hard and brittle and will eventually crack due to vibration or movement in the panel.

Painting Soft Plastics

Some spoilers, such as rear spoilers on Opels or BMWs for example, are made of a very soft plastic and need a different method of painting.

If we sprayed as described above for hard plastics the paint would go on all right but as soon as you pressed it, it would just crack.

The method to use is first, wash it all down, then use a Plastpak primer (which is a grey primer with flexible additive made by ICI) or a similar product from another manufacturer.

Use two coats of this, let if flash off and again do NOT flat it off.

When it is dry, add the top coat of paint. On top of the paint, hardener and thinner the top coats need to have a flexible additive added to it, perhaps as much as 50%. Refer to the manufacturers data sheet.

This additive is a clear material so you will not get the same coverage from the paint. For example, if you needed two or three coats to get good cover without the additive you may need one or two extra coats with the additive. This will give you the flexibility you need for the soft plastic material.

The secret of spraying plastics is to let them dry out by themselves. Do not force them to dry. If you try to force them, every type of plastic has a certain amount of plasticiser in them and the heat will tend to draw out the plasticiser already in the plastic. This looks like little air bubbles coming through, much the same as solvent popping.

If you do get problems with it you have put too much heat on or have not prepared the panel properly. Once you have sprayed the paint on, let it air dry naturally and you should not have any problem.

Cleaning Wheels

Spraying wheels is one of those jobs usually done badly -- with an aerosol can. Follow these steps for a set of well turned out wheels you won't be ashamed of.

First of all, there is no substitute for sand blasting wheels. This will ensure that all the old paint, plus all the old rust is removed. However, there is another alternative which you can try out at home at your own risk.

Get hold of an old oil drum, the 40-gallon size being best. If you can find one cut down until just the lower 12 to 15 inches is left, this will be better.

Put one or two wheels into the barrel. Next add enough water to cover the wheels plus some Caustic Soda solution and heat the barrel until the water is boiling.

This will remove all dirt, grease, oil, paint plus anything else lurking on the surface of the wheels. If necessary stir the water and move the wheels around, taking care not to put yourself at any risk from the boiling caustic solution. Keep onlookers away, especially children and pets, for this is a potentially dangerous process.

When the boiling process is over, rinse the wheels off thoroughly in clean water then dry them off immediately. Assuming you have either 4 or 5 wheels to clean, put another two into the boiler and repeat the process.

They may next require some sanding down with 240 grit paper before they are ready to be sprayed.

How do you spray a wheel? One way to do it might be to fix a spike in a wall and hang each wheel for spraying. Another way is to fix up some sort of wheel stand using pipe fittings. Use some old lengths of conduit piping. It will be useful if you can get a piece about four feet in length, with a right angle bend at the top. Out from that right angle bend you want another foot or so of pipe. It might also be a good idea if this end of the pipe had a joint-type fitting on the end. This will help to keep the wheel in place during spraying. I'm sure you will be able to improve on these ideas.

Masking Wheels and Tyres

If you wish to mask the complete wheel and tyre to prevent it being coloured by over-spray, simple use a plastic dustbin liner over the wheel. Pull it down well over the wheel and tyre until the open edges of the bag are on the ground. If there is a particular reason to do so, then jack the vehicle up, and completely enclose the wheel and tyre in the bag. Then tie the end of the bag shut.

If you wish to spray the wheel without overspraying the tyre, follow these steps:

The recommended way is to deflate the tyre, and insert masking paper between the tyre and the rim. However, this may involve breaking the "bead" by which the tyre seals itself against the wheel rim. When the paper is correctly in place round the tyre, gently inflate the tyre enough to firmly hold the paper in place. You can now spray the wheel. However there is still a problem. You will now spray the wheel nuts as well as the wheel. By the time you have masked off the wheel nuts (do this by simply wrapping some masking tape round them, then over the top of the wheel stud) you would be as well to remove the wheel from the vehicle, remove the tyre and treat the wheels as a separate spraying job.

A tip for masking wheel nuts is to get hold of the little plastic covers which go over wheel nuts on a number of cars. For example my old Ford Cortina estate MK5 has plastic embellishers which simply push over the top of the wheel nuts. Pick up a few on your next visit to a breakers yard or ask at your local dealer.

Spraying over Chrome

Let's imagine you have an old car. It has metal bumpers, on which the chrome is well past its best. Some of it is rusty and pitted. Can you make this presentable by spraying? The answer has to be "perhaps!"

The problem is that chrome is a protective finish, as is paint. Paint needs a surface to adhere to, and chrome exists to be smooth and shiny, therefore adhesion is poor. Your only choice is to rub down the chrome plating to remove it, then treat the resulting metal surface just like any other spraying job.

If the chrome is badly pitted and holed you may be able to prime it and spray it successfully. However the chances are that the paint will flake off over the next few months leaving a mess behind.

Until someone comes up with a special paint designed to adhere to chrome, it is best to assume that you CANNOT paint over chrome successfully.

Commercial Vehicles

This section gives advice on spraying a two-tone commercial vehicle.

Follow the guidance given in the Preparation chapters to remove rust, fill in imperfections and so on. Bare metal, fibre-glass, or aluminium need an etch primer. Acid in the etch eats into the material and gives good bind between base material (be it metal, fibreglass or aluminium) and the primer to be applied. If you use the wrong material you will get flaking. It is essential that you use an etching primer.

Follow up with suitable undercoats relating to the paint system being used. There are two systems to choose from: either use Two-Pack or polyurethane.

After priming and rubbing down, mask off and spray the lighter colour or larger area first.

Allow it to dry for at least eight hours. It can take between eight to twenty-four hours. After this time, check for masking tape marks.

Mask and spray the second colour. When finished, sign writing may be needed.

Commercial vehicles need a good finish as it is the biggest advertisement for the company.

A five year life span should be achieved

with Two-Pack paints; only two years with polyurethane.

Some major manufacturers have developed a two-tone paint system for commercial vehicles where they add a certain additive which draws the paint off, changing it to a base coat and allowing you to spray one or two coats of the one colour. Let it flash off for forty-five minutes to an hour.

Then mask up and recoat the second colour with the same procedure.

Leave that for an hour to flash off.

Lacquering over both materials gives an extra gloss finish as in the base coat system.

Both colours, plus lacquer, all go on in one day saving much more time and money. The lacquers protect pigments in paint against atmospheric conditions and allows gloss to stay much longer.

Two-Pack is being used more and more as the vehicle does not have to be taken off the road every two years. The difference in costs in paint is more than made up by keeping the vehicle on the road for up to five years instead of two years. The vehicle only earns money when it is working!

Other Commercial Systems

Normally, there is no point heating paint. Warming Cellulose or Two-Pack products will achieve nothing. However, there is a heated system more common in commercial vehicle spraying. Used when spraying Synthetic paints, this hot water system is called "hot spraying."

A container of hot water with a thermostatic control is switched on and a tin of paint put into the tin of hot water.

This thins the Synthetic, so no thinner is needed. This system is used on 40-foot tankers for example, as when the paint starts to dry out it thickens back up to cover in one coat.

Sign Writing & Advertising

Signwriting is an art in itself, but is a fast-disappearing skill. You must be a master sign writer if you are to be any good. Nowadays plastic motifs and plastic signs are used as transfers. The signwriter's job is disappearing.

Refer to Chapter Twelve for details of how to apply transfers and other "stick-on" products.

LEFT: An American Ford Mustang. The black coachline could be painted but would require very careful masking. Much easier -- but still needing a lot of care -- are transfers which stick onto the paintwork.

ABOVE: Signwriting or transfers? Either way, it can take several days to prepare all the decorations on this Jordan Formula One racing car.

As with everything else related to vehicle painting, it is attention to detail which makes the difference between a quick "blow-over" respray and a first-class job.

Cellulose - Step By Step

General

The first rule is -- DON'T MIX PAINT SYSTEMS.

Translated into plain English, don't use paint from two different manufacturers on the same job. Order plenty of paint before you start the job and don't swap brands mid-way through a job. If you get into difficulties many paint suppliers have problem-solving technicians who will tell you what you did wrong.

Cellulose is the best type of paint to use for the first time or inexperienced sprayer. It dries quickly, allows mistakes to be corrected easily, and gives a good finish. Polishing will be required to get a good shine, but the time spent achieving this shine costs the amateur nothing.

Initial Preparation

Remove as much trim as possible from the vehicle. By trim we mean - wing mirrors, badges, chrome strips, bumper bars, door handles and so on. It is almost as quick to remove these items as it is to mask them properly.

Wash the car with warm water, but do not use any additives such as shampoo. The idea is to remove all oil, grease, wax and polish coatings from the vehicle.

Dry off the vehicle with clean rags.

Wipe over the areas to be sprayed with a de-greaser. This is to get rid of any remaining oil, grease, wax or polish.

From now on ***DO NOT TOUCH THE VEHICLE WITH BARE FINGERS.*** A small fingerprint could show through all your spraying and mean a panel having to be re-done. Treat the vehicle as being surgically clean and if you must touch something wear cotton gloves.

Flat off the entire area to be painted using wet and dry 300 grade paper. The professionals add washing-up liquid such as Fairy Liquid to the bucket of water. This helps the paint particles to float off the paper and avoids the wet and dry paper becoming clogged. It also allows a smoother passage of the paper across the vehicle's surface. This may be obvious, but it is more comfortable to work with warm water than cold, isn't it?

Dry off the body with compressed air if you have it available. Get the air into all the nooks and crannies where water may be hidden. If no compressed air is available, dry off the body thoroughly with fluff-free rags.

When this rubbing down is completed, begin masking. To do this properly should take several hours. Remember the secret of paint spraying is in the preparation.

When everything is masked off properly, wipe over the prepared surfaces with a tack rag. This will remove any remaining bits of dust, fluff etc.

Cellulose - Step By Step

Spray on a couple of coats of primer.

Attend to any low spots which may have shown up by using spot putty.

Flat down the Cellulose primer ready for the top coat. Use 600 or 800 grade wet and dry paper.

There are two ways of getting a finish with Cellulose paints depending on how much time you want to spend.

The "Quick" Method

If you opt for the quick method, thin down the top coat in the ratio of one, to one and a half, (that is 40% paint 60% thinner). To be free of orange peel you will need some extra coats. Most paint manufacturers recommend one to one, or one to one and a quarter. Based on many years of experience, Trevor sprays at 40% paint to 60% thinner. This way, the coats are much cleaner and free of orange

The final primer coats are being sprayed here. The operator is spraying the rear quarter panel in this shot. Note the heavy masking paper on the wheels, door and windows. On the car roof, plastic sheeting is being used to keep unwanted overspray off the freshly painted panels. Please don't trip over that air hose, Mr Sprayer! Photo: Bert Wiesfeld.

peel. Because you are using more thinner than paint, you need one or two extra coats.

In this quick procedure use four or five coats of Cellulose. Leave to dry, then give a good T-Cut or compound which will give you good gloss.

The "Slower" Method

The other way to spray Cellulose is to apply say, four coats mixed at one to one and a half.

Leave to dry.

Flat with 800 paper.

Next apply another two or three coats mixed in the ratio of 30% paint to 70% thinners.

Leave the Cellulose to dry, then flat with 1500 grade wet and dry paper. Use soap (such as a spot of Fairy Liquid in the water) to stop clogging. This will give a glass-like finish, very similar to a Two-Pack finish.

Polishing is also much easier. Using a polishing machine with a lambswool mop or sponge head will cut down the polishing time by half.

Be careful with a polishing machine as it is easy to cut through the paint on the edges of panels.

If you do, it will mean going back several stages to re-coat the panel.

Masking materials include masking tape (wide and narrow), plastic sheets to cover large areas, and plastic refuse sacks to mask wheels etc. You will also need lots of strong brown paper plus lots of newspaper, which is all right to use as long as you don't use single sheets. That craft knife needs to be really sharp, so buy a new blade before you find out just how blunt the old one is. Masking is an art which needs lots of practice.

Synthetic - Step By Step

Synthetic Paints

Because Synthetic paints are oil-based, you can only apply Synthetic top coats over the top of a Synthetic finish. If you apply Cellulose, One-Pack acrylic or Two-Pack acrylic over Synthetic you will find that the solvents eat through the paint film causing it to craze or pickle.

Choose Synthetic finishes carefully. If in doubt, seek advice.

You may need an isolator if you want to spray a finish ***other than Synthetic*** over Synthetic.

WARNING: Be careful with isolator. If you break through your primer and isolator, back to the Synthetic finish, and put top coats other than Synthetic over, then pickling will occur on the areas broken through

Be careful choosing primers for Synthetics too, as solvents used in thinners for primers tend to attack Synthetic paints.

Initial Preparation

Refer to the instructions at the start of Cellulose Step by Step on page 138.

Synthetics - Step By Step

Flat down the areas to be painted, using 600 grade wet and dry paper.

Spray on two coats of isolator. Leave to dry but do NOT flat.

Next, recoat with whatever primer you have chosen, either Two-Pack or Cellulose primer.

Then recoat top coat over primer with Cellulose, Two-Pack or One-Pack acrylic.

You can get away with putting Cellulose primer over Synthetics if they are applied very, very dry. You can tell when it is dry as there is no shine to the primer as it goes on. This way, solvents get taken out in the air before they reach the paint film.

If you use spray putty put this on very, very dry. You will get away with that as well. Normally you need Synthetic primer or a similar soft solvent primer which won't react.

When you come to spraying top coats of Synthetic you need fewer top coats as you would with Cellulose or One-Pack acrylic as the product is much thicker.

Put on a mist coat all over and let this flash off for about 5 minutes. The mist coat gives the next coat the grip it needs to adhere to the paint to stop it running.

The second coat is a full wet coat. Let it flash for 15 - 20 minutes depending on conditions, then follow it with one last top coat, making it another wet one.

Drying of Synthetic paints depends on the conditions. It is more important to have a good air-flow rather than heat, mainly because air flow dries out oil better than heat. You need both though, good extraction and good flow of heat.

When thinning paint whether it is Cellulose, One-Pack or Two-Pack, stick to the appropriate thinners for the paint.

Lots of people use cheap economy thinners. You are fooling yourself by using these cheap thinners. They save you a couple of pounds in the beginning but you need a lot more (and harder!) work to polish up.

Regarding Synthetic thinners, in the trade cheap Cellulose thinners are put in with Synthetic paints. This does work in that it flashes off Synthetic a bit quicker than an oil-based Synthetic thinner would, but it lowers the gloss in which case you are just fooling yourself again.

If you are good enough with the spray-gun you should not need to polish a Synthetic finish -- it should give you gloss straight from the spray-gun.

If you have to polish, make sure the polish is ammonia-free.

If you use polish such as cutting agent

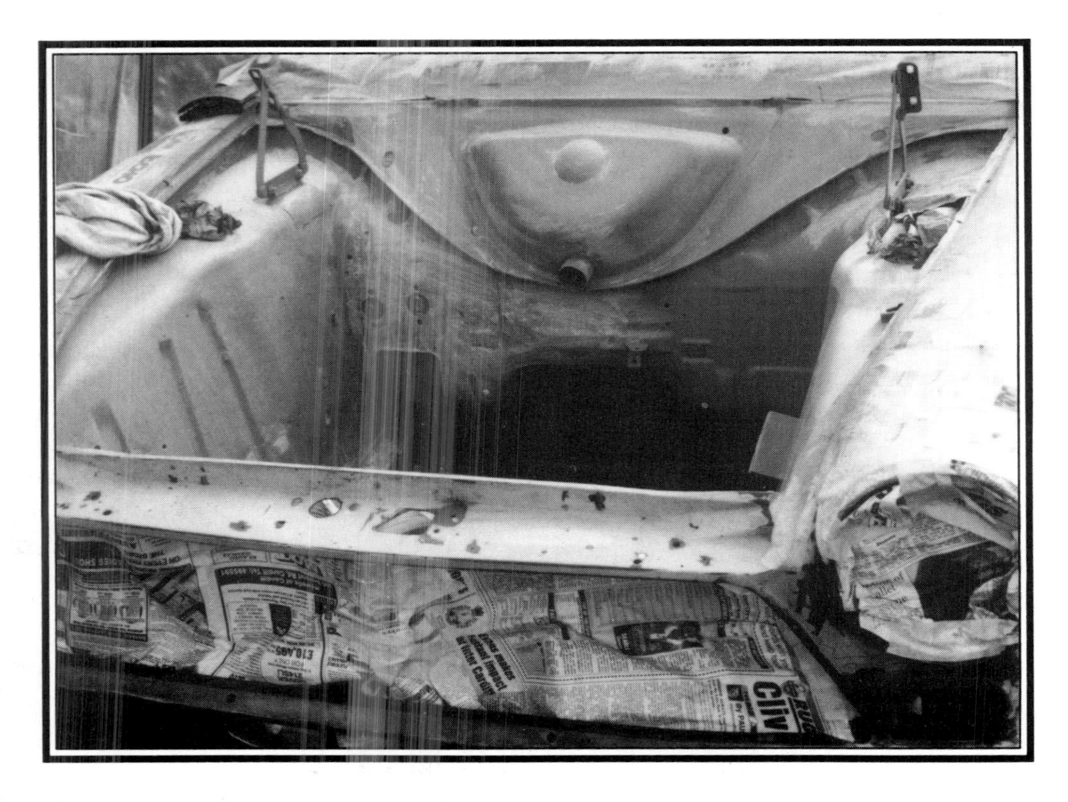

When you use spray putty make sure you allow plenty of time for each coat to flash off before spraying another. If you don't, then the top coat may be dry but underneath may still be wet. When the coats underneath do dry out, the top coats may crack which means a lot more work to put right. As Trevor says, never rush a job.

compound or T-Cut with ammonia, then when you polish the surface it will take gloss off slightly. If you use lots of ammonia it will take gloss away completely.

More on Spray Putty

Going back to spray putty, they are not bad if used correctly. In the wrong hands or used incorrectly they can become a nightmare.

The problem is there is no thinner applied with spray putty -- it is straight out of the pot onto the panel. This means it is quite a thick material.

Put on a coat of spray putty and let it flash off for at least 5 minutes. This time will allow it to dry and not trap any solvents or air between coats.

Some operators use this product incorrectly in that they put on three, four or even five coats in next to no time. They let it dry for 2 hours. The top coat dries sufficiently to flat down ready for top coats, but because such a lot of paint has gone on in such a short time the top coats are dry and underneath it is still drying out and moving.

You will find in a couple of days that the underneath has dried but that cracks will appear in the top surface. It is essential that you let spray putty dry through. But really this applies to all forms of painting.

There is no quick way to dry a paint job.

If Synthetic paint says 8 - 24 hours drying time on the can, then it needs that time. To force dry it will force dry just the surface, resulting in underneath coats still being wet.

Basecoat & Clear - Step By Step

Basecoat and Clear

Paint used in the refinishing trade are Cellulose Basecoat and Clear, and Two-Pack finishes which include Basecoat and Clear.

Two-Pack are the paints of the future. They give gloss from the gun and offer a lifespan of 5 years against the weather, so more and more garages are turning to Two-Packs. Ideal conditions for Two-Pack are a professional spray booth but you can get away with spraying in lesser surroundings if you have an air-fed breathing system. This is a MUST!

Two-Pack systems, as far as Basecoat and Clear are concerned, are moving across to solid colours as more and more manufacturers are using solid Basecoats with coat of lacquer over top. They find that with a coat of lacquer over the top the paint does not fade and the weather does not attack paint pigments too much.

Initial Preparation

Refer to the instructions at the start of Cellulose Step by Step on page 138.

Step By Step

Once you have the primer applied, as in all the other paint systems we've been looking at, proceed to put top coat on.

Flat down the primer with 800 wet and dry paper, or 1000 wet and dry paper. Some manufacturers recommend 800 some say 1000. This is entirely up to you.

When flatting is complete apply Basecoat in the ratio of one to one, (one part paint to one part thinner). Apply three coats of Basecoat with about 5 minutes between the coats. This time can be reduced if you have good air-flow and heat in the workshop.

Coats of Lacquer on top of Basecoat

Panel being sprayed

A clear coat of lacquer is sprayed over the top of the basecoat to provide protection from the weather.

Once the Basecoats are laid down nicely, (you need aluminium laydown techniques, as no gloss is required), the Basecoat dries in a semi-matt finish. It is not essential to put paint on wet. Medium coats will do.

Once you have finished spraying the panel with the appropriate number of Basecoats, wait some 20 minutes for flash-off before applying the Clear coat.

Clear coat is available in two different systems. Cover the Basecoat with One-Pack Clear coat straight from the tin. No thinners are required but polishing may be needed.

The Two-Pack system contains Isocyanate and are usually mixed in the ratio of two parts Clear coat to one part hardener.

Apply two or three coats of Clear coat, or as recommended by manufacturer. With this system you'll get gloss straight from the gun.

If you want high gloss with One-Pack Acrylic clear lacquer, put four coats on but be prepared to give a really good T-Cut. When you've polished One-Pack up it will come very close to a Two-Pack finish.

Metallics - Step by Step

Metallic Spraying Problems

Metallic paints contain aluminium particles which reflect the light to give a shiny appearance.

Metallic paints are by far the most difficult paints to spray.

Striping is the most common problem encountered when spraying Metallics. This is generally caused by one of three things:

1) Insufficient overlap between coats leading to a "dry edge", (slow thinners will overcome a dry edge), or

2) Using cheap thinners which flash-off at the wrong speed for the paint, or

3) Incorrect spray-gun technique. This third category includes the gun being too close to the panel, the gun width being too narrow, or the gun at the wrong angle to the panel.

Metallic paints need a thinner which allows the Metallic particles in the paint to settle before flash-off. If the flash-off is too rapid, then the Metallic particles will not spread properly, leading to areas of higher concentration of particles and patchy areas of low concentration.

Metallic paints change colour depending on the wetness or dryness of the spray.

A **wet spray** means metal particles tend to sink to the bottom, hence a **darker finish.**

A **dry spray** means the metal particles have no chance to sink, therefore they are closer to the surface and can create a **lighter finish.**

Metallics are probably the hardest product of the lot for anyone to spray. To get aluminium movement (to lay down properly) plus getting a gloss is a very difficult task.

Initial Preparation

Refer to the instructions at the start of Cellulose Step by Step on page 138.

Metallics - Step By Step

When spraying Metallics there are two ways to do it:

First Metallic Method

Apply quite a heavy coat, followed by a medium coat, followed by another medium coat then a dusted out coat. The dusted coat must be applied straight after the previous coat.

(Dusted out means hold the gun further away from the painting area and dust and spread the aluminium over the painted area. This gives good laydown and an even surface). The problem with Metallics is if you try to get a good laydown and pile the paint on too wet, you get a striping process, that is, light and dark areas.

The dark areas are where aluminium is buried by a heavy coat, and light areas where paint is drying out quicker than heavier areas and aluminium is coming to the surface.

Second Metallic Method

The second way to spray Metallics, and the way Trevor does it, is to go for aluminium movement rather than gloss level. Do this by spraying medium coats and dust away at the aluminium, followed by polishing the surface when dry. This can be done with One-Pack and Cellulose.

For Synthetics you need to dust away at them. Trevor puts on a coat of blending clear mixed with some Synthetic Metallic paint already used as a last coat, giving you a good flow.

On Two-Pack Metallics he cheats! Trevor puts medium coats on followed by one coat of lacquer. The one coat of lacquer dries into the paint and gives you high gloss Metallic. Two coats on gives you a fairly false finish as gloss is too high,

Metallic finishes are tending to be phased out and replaced by Basecoat and Clears. The clear lacquer helps fight the elements and so protects the pigments.

Continued on page 147.

Metallics - Step By Step

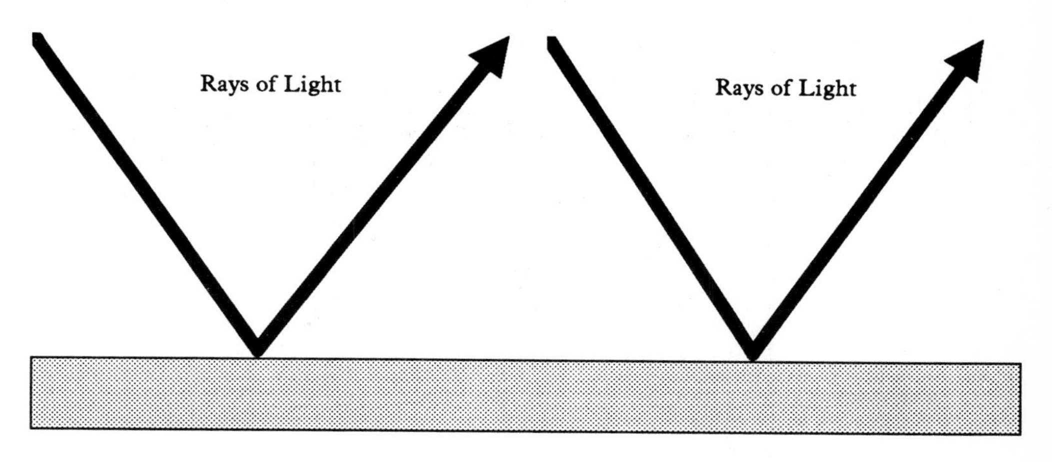

On a smooth surface such as a mirror or a polished paint finish, light rays are reflected from the surface. The eye receives the reflected rays of light and informs the brain that the surface is "shiny". A dull (or non-shiny) surface would absorb more light than it reflected.

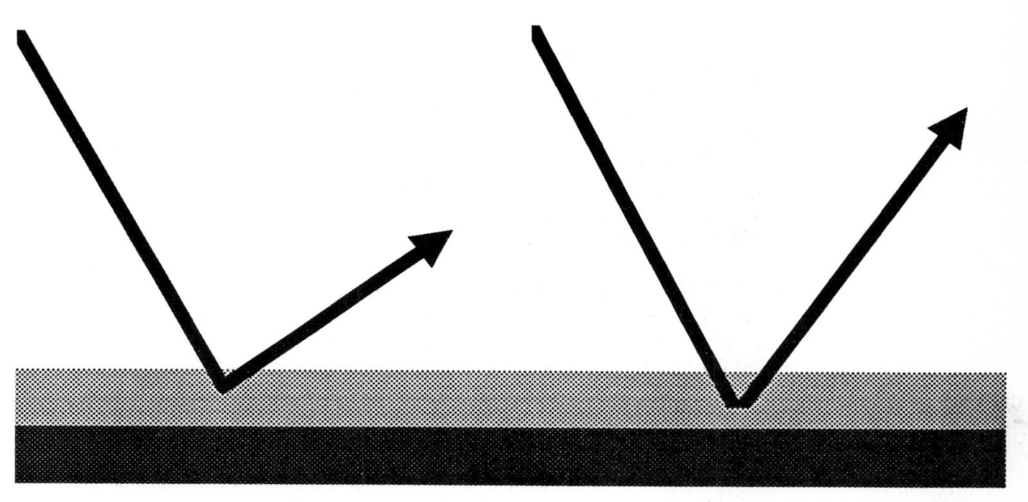

On a "wet" Metallic finish, where most of the aluminium has sunk, the eye will see a darker finish as shown above.

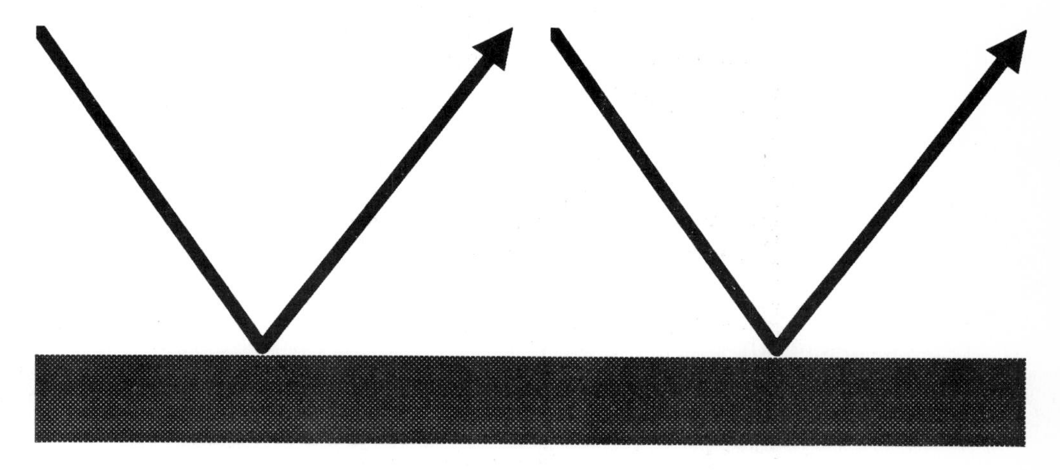

On a "dry" Metallic finish where the aluminium has not had a chance to sink, the eye will see a lighter, more "shiny" reflection.

Metallic Manipulation

The finished appearance of Metallic paints can be made darker or lighter, depending on certain factors often described as "variables". These variables are: spray booth conditions, spray-gun settings, thinners, and perhaps most importantly the spraying technique used.

These variables are illustrated in the following table.

Variable	To Lighten Colour	To Darken Colour
Spray Booth		
Temperature	Increase	Decrease
Humidity	Decrease	Increase
Extraction/Airflow	Increase	Decrease
Spray-Gun		
Needle Control	Close	Open
Fan Width	Wide	Narrow
Air Pressure	Increase	Decrease
(See NOTE 1)		
Thinners		
Type of Thinner	Fast†	Slow
Amount/ratio	Increase	Decrease
(SEE NOTE 2)		
Technique		
Gun Distance	Increase	Decrease
Gun Speed	Increase	Decrease
Flash Off Time	Increase	Decrease

Note 1: If alternative fluid tips are available, fit a smaller fluid tip to lighten the colour and a larger tip to darken the colour. If alternative air caps are available, fit a high air consumption cap to lighten the colour and a low air consumption cap to darken the colour.

† Trevor advises that using a fast thinner will stop the aluminium flowing. He suggests a slow thinner and allow longer to dry.

Note 2: Retarders can be added to the thinners at up to 10% ratio to darken the colour. However, do not use any retarders if you wish to lighten the colour. Select another variable to alter.

Blending Metallics

On a partial respray this is very similar to the Step by Step process outlined in the next Chapter for Mica.

For example, if you have a four-door saloon car and you want to paint two doors on one side.

Repair if required, prepare the paint surface, priming if necessary. Flat down the primer ready for the next stage.

Where you are repairing scratches, for example, you are going to have to fill the scratches, prime and flat down. Very light scratches may just need rubbed down and dusted with primer.

If you have to prime both doors in full then polish front and back wings with T-Cut and mask the vehicle off as though you were going to respray the complete side.

Scotchbrite the front and rear wings with superfine Scotchbrite. Then wipe down all the surfaces with a de-greaser.

Next wipe all surfaces with a tack rag.

Strain the paint into the gun and spray some coats on.

The first coat must cover both doors and just slightly over the edges of front and rear wings.

The second coat covers both doors and a little bit further out over the edges of front and rear wings.

The third coat (which should be the final coat if all has gone well) is spraying the two doors and a little bit further out over the edges of front and rear wings.

Next, thin the paint, lower the air pressure and arc the gun to fade out. See the illustrations on pages 128 and 129.

Metallics are very, very difficult, especially if they are single layer Metallics.

You need to fade out your edges by using a thinner to fade out with. Some paint manufacturers have a special fade-out thinner for this purpose, others use ordinary thinners. The idea is to soften up the edge of the paint underneath to accept the new paint.

If you want to do a very professional job, Trevor recommends you lacquer across your panels and then it will fade-in fairly easily.

There is also a product on the market called a "blending clear" which blends your colour out for you. If you are spraying a Metallic you put your first two colour coats on, then add a little bit of blending clear (refer to the manufacturers instructions) to the third coat.

Alternatively, add a little bit of blending clear to each coat but add a little more each time.

In the final coat you need approximately two thirds blending clear to one third paint. Add a little bit more thinners and your fade-out technique should go well.

With fade-out techniques don't think the job is complete if you can still see the edge. If you look at the job and can hardly see the edge, perhaps just a little bit of dryness which will come out with polish, you will be okay.

But if you can still see a difference or variation in colour, your fade-out technique has not worked. You will have to carry on fading out, or flat down and start again. This is a very difficult technique to master, but if you use lacquer it makes things so much easier.

Two-Pack - Step By Step

YOU MUST USE AIR-FED BREATHING APPARATUS WHEN SPRAYING TWO-PACK MATERIALS WHICH CONTAIN ISOCYANATES.

General

Two-Pack paint materials consist of paint, hardener and thinner. Consult the paint manufacturers' data sheets, available when you buy your paints.

Two-Pack paints are mixed (approximately) in the ratios ten parts colour, three parts activator, three parts thinner. This works out at about 13 parts paint to 3 parts thinners. This means that for a given vehicle you use much less paint and thinners to achieve an even better gloss. The cost is in pound notes, currently around £30 per litre for paint, plus hardener, plus thinner, (you ***can*** get it cheaper!), plus the health hazards of Two-Pack, plus the drying time.

Problems also take much longer to sort out -- if you get a run you must wait until it is dry before re-working the area.

Initial Preparation

Refer to the instructions at the start of Cellulose Step by Step on page 138.

Primers

At the primer stage look to Two-Pack primers and hardeners. They don't sink like Cellulose or spray putties do. Use a hardener which accelerates drying time. This is an Isocyanate hardener, so air-fed breathing apparatus must be used to safeguard your health.

These Two-Pack primers go off very, very hard. Once they are fully cured, flat down with 800 wet and dry. Don't forget drying with the airline, degreasing and wiping with a tak-rag before applying two or three coats of top colour. The gloss from the gun shouldn't need polishing.

When you apply Two-Pack materials the drying process relates to hardeners supplied with paints. There are hardeners for air drying or for stoving facilities. Remember to specify the correct hardeners for your workshop conditions. Remember that times stated are for fairly good workshop conditions. If your workshop does not meet the stated conditions the drying process may be slowed, especially where dampness or coldness are concerned. These stop the chemical reaction taking place as quickly as it should.

Mica Finishes

Mica finishes are Base coats with a little bit of special pearl material added into the mix formulae.

Unlike Pearl coat, Mica can be blown out or faded in, similar to a Basecoat system.

Two-Pack - Step By Step

Repair and prepare the required area, prime it. For example if you are spraying a door, both of the adjacent panels must be polished with T-Cut to get the colour back. Then Scotchbrite the areas to be lacquered over. This is very, very important and you must Scotchbrite thoroughly, otherwise you will not get the adhesion and the lacquer will peel.

Mask off the areas to be sprayed. Spray the door, fading in to the adjoining panels. If you spray, say three coats of colour, the first coat of colour takes in the door and takes in a little bit on the adjoining panels. The second

The final Two-Pack top coats are being sprayed here. You can see the level of gloss being achieved by the reflections from the lights. The operator is spraying the engine bay in this shot. Note the masking on the wheels, door and windows. The car is a Ford 100E which was found in a stable and restored and painted by Bert Wiesfeld of the Netherlands. Photo: Bert Wiesfeld.

coat takes in a little bit more on the adjoining panels. The third coat is then similar, coating the door but going further down the wings. Now add a bit more thinner and reduce air pressure to "fade-out."

Fading Out

When spraying normally, keep the gun parallel to the panel being sprayed. When fading out, you arc the gun without triggering off, to mist some paint material down the edges of the panels. If you used the normal parallel sequence you will probably find you get a silver edge, as the aluminium has nowhere to flow. Finally, lacquer the panel.

Pearl Coat Finishes

Pearl Coat is a three finish system whereby we find the colour the vehicle was sprayed in, find the colour code, for example a Renault colour code may say pearl white. We need to associate Pearl White with the proper ground coat which goes underneath it. The "ground coat" is a special undercoat which is also the main colour. It is otherwise known as a background coat.

Pearl Coat cannot be faded out, so we need to find the proper ground coat to give us a colour match, edge to edge. Take the ground coat and add a basecoat thinner to it and spray two to three coats of colour on to give us the base finish. Next mix the Pearl Coat up with some thinners and spray that. It does not matter how many coats -- three to five coats -- as it does not cover anyway, it just gives the Pearl effect. Leave that to flash off. Then spray lacquer with hardener added to give the final finish.

It is most important to find the proper ground coat. For example, people have used Ford Diamond White or another basic white and you don't get the same effect. You must have the correct base coat colour underneath your Pearl to give you the correct colour match, otherwise you will be in trouble!

Finishing & Polishing

Getting Good Gloss

Whilst it is entirely possible that you will achieve a good gloss from Cellulose straight from the spray gun, you can enhance the gloss considerably by polishing the paint. There are several ways to do this. Remember, this is the stage of the job where you wish you had spent a few more pounds on better quality thinners!

Polishing procedures are described below. Remember that it could take from four to six ***hours*** to polish a Cellulose respray to a high gloss. If you were paying a professional respray shop to do the work for you they would charge you for this polishing time (or take short-cuts). One way for them to get a better gloss straight from the gun is to use a Two-Pack paint. This requires little polishing.

Mere mortals like you and I *(writes Tommy)* are going to have to polish our new paint job, either by hand or with a polishing machine!

Cellulose and Clear

Cellulose can be covered with a coat of clear lacquer. But if you spray lacquer straight over the top of Cellulose when the paint is still wet it will suck into the underlying coats. This is due to the fact that when Cellulose is drying it tends to suck into itself and dull off slightly.

To get over this let the Cellulose finish harden off properly, then polish with T-Cut or cutting compound, to achieve a really high gloss.

Then go over the surface with a Scotchbrite pad, to scuff it to give the surface a grip for the lacquer.

If you do not scuff it up, lacquers will just peel off, or if you use too coarse a paper the lacquer will go over the top and show scratch marks.

Use the Scotchbrite pad to scuff the surface up very, very lightly and the scratches you are putting in will be covered by the clear lacquer.

There are two lacquer systems to choose from:

Either use One-Pack Acrylic lacquer over the top and leave it to dry, then polish the lacquer up to a finish,

OR

Use the Two-Pack system and give it two coats of lacquer. This gives you a good gloss straight from the gun.

Remember, Two-Pack gives off Isocyanate, so an air-fed mask is necessary, plus an air filtration system between compressor and mask. This filtering system will take out any dirt present in the air line.

Flat and Polish

If you do not wish to cover Cellulose with lacquer then you may want to flat and polish the Cellulose.

This technique involves flatting down the finished Cellulose top coat with a 1200 grade wet and dry paper. It also helps to use soap or washing up liquid to stop this very fine paper from clogging. Carefully flat the paint, trying to avoid inflicting any deep scratches on the panel.

When this is complete, and it will take several hours to flat an entire car, you are ready to polish.

Polish

This polishing stage involves using T-Cut and a number of clean, lint-free soft cloths. These clean soft cloths are important. If they are not clean you are going to transfer any contaminants onto the panel, and if they are not soft you are going to scratch your hard-won painted surface.

Trevor recommends you use a Farecla

When you have polished your paint to a mirror-like shine, how can you judge how shiny it actually is? One way is to look at the reflection. We polished up a small area of neglected Ford. Six years of dirt and grime vanished and you can see the T-Cut tin and white clouds reflected.

compound first, then use T-Cut you will find you will do the job in less time. Farecla is available in various grades of coarseness, so be sure to ask your dealer's advice when buying. Use the Farecla to do the hard work, then switch to T-Cut.

Work on one panel at a time, working the T-Cut over the area selected until it has almost disappeared. Then add more T-Cut and repeat the process. You may wish to continue this process two, three, or more times.

When you have finished working in the T-Cut, it is time to polish off the entire treated area with another soft clean cloth. Be especially careful not to clog up this polishing cloth with T-Cut. If the cloth shows signs of becoming clogged up with paint, throw it away and use another cloth.

If you don't fancy spending hours polishing a car by hand, you may want to consider what mechanical aids can be used. Whether air-powered, or electrically powered, there are two criteria for selecting a polishing tool. It must run slowly, (high speed polishers are OUT) and the mop must be soft. If you fit the wrong type of polishing mop you run the risk of scouring and scratching the whole job.

A lambs-wool mop is usually recommended for polishing.

After the T-Cut stage, polish your panels with a suitable polish. Here, Trevor recommends Autoglym as being one of the best waxes on the market.

The more time and effort put into the process, the better the finish will be. Cellulose paint will need to be polished now and again to keep the gloss. If you leave it more than 12 - 18 months then the paint surface has been neglected. It will be dull and flat and will require another arm-aching T-Cut session to bring it back to life.

Other types of paint will give you a better gloss, straight from the gun. Metallics will give a better gloss than Cellulose, while the best gloss of all will come from Two-Pack paints. Remember the health risks associated with Two-Pack paints.

The "Watch" Test

Having polished your panels for hours, until your arms ache, how can you test the shine you have produced? One way is to look at the reflection made by your watch. A dull panel with no shine will probably not reflect much at all. A mediocre panel with some shine will reflect the outline of your wrist and the watch.

But look at a highly polished area. The watch is reflected, you can even tell the time in the reflection. Individual hairs can even be seen on your arm. That is the sort of mirror-like shine which can be achieved!

I suppose the more technical readers will be able to devise suitable electronic shine or reflection measurements. You may be able to use something like an exposure meter used with older cameras. However, the good old watch test is free and almost as good.

This test is also referred to in some books as the coin test when a coin is used instead of a watch.

Dealing with Overspray

If you have masked properly, then there should be no overspray on other panels, wheels and tyres, chrome, glass, or glass rubbers.

However, we are only human, so if there is some overspray how do you deal with it?

Dealing with glass first, the best way is to carefully scrape off the paint with a razor blade or craft knife. If you are cleaning paint off rubber this way, be very careful not to cut the rubber.

For overspray on metal parts first try a thinners-soaked rag. This should remove most of the problem.

Remember prevention is easier than cure where overspray is concerned.

Air Brushing

Air brushing is the technique of painting small details or decorations onto a vehicle. It can be small flames or waves, or a large mural covering the entire side of a van.

Firstly, you don't need much air pressure on an air brush, which is basically like a pen.

There are several different types of air brushes on the market - you can get aerosol tins which go on the end of the air brush. You can buy small compressors suitable for air brushing which will probably cost under £100. These would be ideal for someone really keen to develop their air brushing skills.

You can also buy a reducer to cut down the pressure if using an air brush from a full size compressor.

You make you own design, such as country-side scenes, dice, jigsaw patterns, flames, and so on, but there are several books available to help you.

Put simply, you mask off, spray an area, then mask off again. If you are spraying a large area, you can use your normal gun with reduced pressure and fading out techniques. Then use the air brush for the smaller details.

To get clouds, or waves for example, cut out a transfer (or stencil) first then fix the stencil on with masking tape, and spray the area.

With air brushing you are limited only by your imagination.

Applying Coach Lines.

This section deals with the stick-on coach lines which can be purchased from most motor factors or paint suppliers.

Applying transfers (known as decals in the USA) or coach lines is a job best tackled by two people. First of all, polish the areas where the stripe will be applied with T-Cut.

T-Cut both sides of the car, assuming a stripe is going to be added to both sides of the vehicle. If you only T-Cut one area and get the full colour back, then it is going to look a bit odd against the other, duller, panels.

Wipe the areas down with a de-greaser to make sure there is no polish left on the panels, otherwise the coach lines will not stick.

Most coach lines come with a clear backing paper which needs to be peeled off. Take this off, then apply the tape onto the panel.

Press down one end of the coach line then line up with the eye to the other end of the vehicle. (Don't do one panel at a time -- do the whole line on one side of the vehicle!).

Do NOT rub it down inch by inch as all you will get is a line which moves up and down, all along the panel. You will get a "bent" straight line!

If you have someone to help you, then get them to hold one end firmly, while you reel off the tape from the roll and line up by eye. Then get your assistant to come along the line of tape and smooth it over very lightly.

Then stand back to make a final check for accuracy. If all is well, go over the line with a soft cloth and push down the tape all along. Then go back to the start of the line and pull off the clear tape, away from the stripe. This leaves the stripe on the vehicle and the clear tape ready for disposal.

Go over the line again with the soft cloth and make sure the tape is pushed down firmly.

Then you need a sharp craft knife, such as a Stanley knife, to trim the edges where the doors open, for example. It goes without saying that you should not open any doors until the coach line has been trimmed with the knife, otherwise you will tear it all from the panels.

Applying Transfers

This section deals with applying transfers, such as logos, lettering or other decorative, stick on products.

There are two forms, a water-based one and a dry one. The dry one is very like applying a coach line as described above. It has a backing paper which you pull off and throw away, while the actual transfer is then stuck to the vehicle.

With this type you need to start at one edge of the stuck on transfer and smooth all the air bubbles out. Use a smooth piece of plastic such as the plastic spreader you get with body filler.

Work from one edge and keep smoothing down until you work all the air bubbles out. The end result should be a nice, flat transfer with no wrinkles, folds, or bubbles.

With the water based transfer, you need soap and water on the vehicle first as your backing. Then put your transfer on. The water will activate the adhesive just like licking a stamp before sticking it on a letter.

With this type of transfer you will probably find that you still get air bubbles trapped underneath the transfer.

Use the smooth plastic spreader as described in the previous paragraph. Try to work out most of the air bubbles, but be prepared for a few stubborn ones to remain.

Most of these will dry out but if some remain after a reasonable time the best thing to do is go around with a pin and prick the air bubble to squeeze the air out. They should then go back down flat.

Air Line Pressure Drop - Pounds per Square Inch (psi)

Length of air line in feet					
Feet	10	15	20	25	50 feet
1/4 inch internal diameter air line					
40 psi	8	9.5	11	12.75	24
50 psi	10	12	14	16	28
60 psi	12.5	14.5	16.75	19	31
70 psi	14.5	17	19.5	22.5	34
80 psi	16.5	19.5	22.5	25.5	37
90 psi	18.75	22	25.25	29	39.5
5/16 inch internal diameter air line					
40 psi	2.75	3.25	3.5	4	8.5
50 psi	3.5	4	4.5	5	10
60 psi	4.5	5	5.5	6	11.5
70 psi	5.25	6	6.75	7.25	13
80 psi	6.25	7	8	8.75	14.5
90 psi	7.5	8.5	9.5	10.5	16

Air Line Pressure Drop - Bar (Barometers)

Length of air line in metres

Metres	3	6	9	12	15
Internal diameter 6.35mm					
3 Bar	0.58	0.82	1.08	1.4	1.72
3.5 Bar	0.68	0.96	1.24	1.58	1.92
4 Bar	0.8	1.08	1.4	1.72	2.07
4.5 Bar	0.92	1.24	1.56	1.88	2.22
5 Bar	1.04	1.40	1.74	2.06	2.38
6 Bar	1.23	1.70	2.06	2.36	2.66
Internal diameter 7.94mm					
3 Bar	0.20	0.27	0.37	0.48	0.61
3.5 Bar	0.25	0.32	0.41	0.55	0.69
4 Bar	0.29	0.36	0.47	0.61	0.76
5 Bar	0.37	0.47	0.60	0.75	0.91
6 Bar	0.49	0.62	0.75	0.90	1.16

NOTE: 1 BAR = 14.2 Lbs per square inch.

Air Pressure Conversion

Pounds per Square Inch to Kilogrammes per square centimetre.

10 psi = 0.7 Kg/square centimetre,

20 psi = 1.4 Kg/square centimetre,

30 psi = 2.1 Kg/square centimetre,

40 psi = 2.8 Kg/square centimetre,

50 psi = 3.5 Kg/square centimetre,

60 psi = 4.2 Kg/square centimetre,

70 psi = 4.9 Kg/square centimetre,

80 psi = 5.6 Kg/square centimetre.

Bars to Pounds per Square Inch

1 bar = 14.2 lbs/sq inch,

2 bar = 28.4 lbs/sq inch,

3 bar = 42.7 lbs/ sq inch,

4 bar = 56.9 lbs/sq inch,

5 bar = 71.1 lbs/ sq inch,

6 bar = 85.3 lbs/sq inch.

Temperature Conversion Table

0° C = 32 ° F

10° C = 50° F

20° C = 68° F

30° C = 86° F

40° C = 104° F

50° C = 122° F

60° C = 140° F

Volume Conversions

Pints to Litres

1 pint = 0.568 litres,

2 pints = 1.13 litres,

3 pints = 1.71 litres,

4 pints = 2.27 litres,

5 pints = 2.84 litres,

6 pints = 3.41 litres,

7 pints = 3.99 litres,

8 pints = 4.55 litres.

Litres to Pints

1 litre = 1.76 pints,

2 litres = 3.52 pints,

3 litres = 5.28 pints,

4 litres = 7.04 pints,

5 litres = 8.80 pints,

6 litres = 10.56 pints,

7 litres = 12.32 pints,

8 litres = 14.1 pints.

To convert litres to gallons multiply litres by 0.2205.

To convert gallons to litres multiply gallons by 4.536.

Thinning Ratios

25% = 1 part thinner to 4 paint,

33% = 1 part thinner to 3 paint,

50% = 1 part thinner to 2 paint,

75% = 3 parts thinner to 4 paint,

100% = 1 part thinner to 1 part paint,

150% = 3 parts thinner to 2 paint,

200% = 2 parts thinner to 1 paint,

250% = 5 parts thinner to 2 paint,

300% = 3 parts thinner to 1 paint.

Addresses

DeVilbiss Automotive Refinishing Products, Ringwood Road, Bournemouth, BH11 9LH, United Kingdom

Phone: 0202 571111

Fax: 0202 573488

Manufacturers of paint spraying equipment.

Binks-Bullows Limited, Brownhills, Walsall, West Midlands, WS8 7HW United Kingdom

Phone: 0543 372571

Fax: 0543 360702

Manufacturers of paint spraying equipment.

Spray and Air Systems Ltd, 83 Springvale Industrial Estate, Cwmbran, Gwent, NP44 5BE, United Kingdom.

Phone: 06333 69195

Fax: 06333 2645

Suppliers of compressors and paint spraying equipment.

Health and Safety Executive, Library and Information Services, Broad Lane, SHEFFIELD, S3 7HQ

Phone: 0742 752539

Health and Safety Executive, Library and Information Services, Baynards House, 1 Chepstow Place, Westbourne Grove, LONDON, W2 4TF

Phone: 071 221 0870

Machine Mart, 211 Lower Parliament Street, Nottingham, NG1 1GN United Kingdom,

Phone: 0602 411200

Suppliers of Compressors and other tools. Branches throughout the UK.

Gramos Chemicals International Ltd, Dundas Lane, Portsmouth, Hampshire, PO3 5NT, United Kingdom

Phone: 0705 667421

Fax: 0705 663292

Manufacturers and suppliers of one-piece painting overalls, and other supplies.

Hammerite Products Ltd, Prudhoe, Northumberland, NE42 6LP, United Kingdom

Phone: 0661 830000

Fax: 0661 835760

Manufacturers and suppliers of Hammerite and Smoothrite paints, and Kurust, Joy, Waxoyl and Hermetite products.

Bibliography

The following is not intended to be a definitive list. Some of these books are now out of print, and only available from second-hand bookshops or autojumbles. Where a book is known to be out of print, this has been indicated.

"Panel Craft" by Tommy Sandham, published by Willow Publishing (Magor), Barecroft Common, Magor, Newport, Gwent, NP6 3EB, United Kingdom. Contains a large section on MK1 Cortina body restoration.

"Restoring Small Fords" by Tommy Sandham, published by Willow Publishing (Magor), address above. Contains a large section on small Ford (from 1959 to 1968) body restoration.

"How to Restore Chassis and Monocoque Bodywork" by Tommy Sandham. Part of Osprey's Restoration Series. Now out of print.

"How to Restore Paintwork" by Miles Wilkins. Part of Osprey's Restoration Series.

"How to Restore Sheet Metal Bodywork" by Bob Smith. Part of Osprey's Restoration Series.

"How to Restore Fibreglass Bodywork" by Miles Wilkins. Part of Osprey's Restoration Series.

"Vehicle Refinishing" by North, published by Osprey.

"How to Paint Your Car" by David H. Jacobs, published by Motorbooks International.

"How to Custom Paint" by David H. Jacobs, published by Motorbooks International.

"How to Repair & Restore Bodywork" by David H. Jacobs, published by Motorbooks International.

"The Car Bodywork Repair Manual" by Lindsay Porter, published by Haynes Publishing.

"Basic Bodywork & Painting" by Petersen Publishing Company.

"Basic Painting Tips & Techniques" compiled from Hot Rod Magazine.

"Basic Bodywork Tips & Techniques" compiled from Hot Rod Magazine.

"Metal Fabricator's Handbook/Race & Custom Car" by Ron Fournier, published by HP Books.

"Sheet Metal Handbook" by Ron & Sue Fournier, published by HP Books.

"Practical Classics & Car Restorer on Panel Beating & Paint Refinishing" published by Kelsey Publishing.

"Paint & Body Handbook" by Don Taylor & Larry Hofer, published by HP Books.

"The Repair of Vehicle Bodies" by A. Robinson. Published on behalf of the Vehicle Builders' and Repairers' Association, by Heineman Educational Books.

"Principles and Practice of Vehicle Body Repair" by J. Fairbrother, published by Stanley Thornes.

"Autobody Repairing and Repainting" by Bill Toboldt, published by Goodheart-Willcox.

"Panel Beating and Car Restoration" by Donald Wait, published by Orbis Publishing.

"Automotive Body and Fender Repairs" by C.E. Packer, published by Goodheart-Willcox.

"Automotive Collision Work" by Edward D. Spicer, published by American Technical Society.

"Panel Beating and Body Repairing" by Donald Wait, published by Ure Smith.

If you are interested in art & design or air brushing, contact a specialist supplier such as Graphic Books International of Guernsey, Channel Islands.

Index

Index